Semiconductor Materials and Modelling for Solar Cells

Z. Pezeshki[1]*, A. Zekry[2]

[1]Faculty of Electrical and Robotic Engineering, Shahrood University of Technology, Shahrood, Iran

[2]Faculty of Engineering, Department of Electronics and Electrical Communication Engineering, Ain Shams University, Cairo, Egypt

* tejaratemrooz@gmail.com

Edited by

Inamuddin

Department of Applied Chemistry, Zakir Husain College of Engineering and Technology, Faculty of Engineering and Technology, Aligarh Muslim University, Aligarh-202 002, India

Published by **Materials Research Forum LLC**
Millersville, PA 17551, USA

Published as part of the book series
Materials Research Foundations
Volume 104 (2021)
ISSN 2471-8890 (Print)
ISSN 2471-8904 (Online)

Print ISBN 978-1-64490-142-7
ePDF ISBN 978-1-64490-143-4

Distributed worldwide by

Materials Research Forum LLC
105 Springdale Lane
Millersville, PA 17551
USA
http://www.mrforum.com

Printed in the United States of America
10 9 8 7 6 5 4 3 2 1

Table of Contents

Abbreviations

A

AC: Alternative Current
MOS: Antireflection Metal-Oxide Semiconductor
AFM: Atomic Force Microscopy
a-IGZO: Amorphous a-InGaZnO$_4$
AR: Antireflection
AS: Amorphous Semiconductor
a-Si: Amorphous Si
a-Si:H: Hydrogenated Amorphous Si
AZO: Aluminum Zinc Oxide

B

BPC: Butyl Pyridinium Chloride
BTO: BiTiO$_3$

C

C-AFM: Conductance AFM
CB: Conduction Band
CC: Charge Carrier
CdTe: Cadmium Telluride
CID: Copper Indium Diselenide
CIGS: Copper Indium Gallium Selenide
CIS: Conductor-Insulator Semiconductor
CISO: Conductor-Insulator Semiconductor Organic
CPS: Conductor Polymer Semiconductor
CTS: Copper-Tin Sulfide
CVD: Chemical Vapor Deposition
CZBiTS: Cu$_2$Zn1-xBi$_x$SnS$_4$

D

DCNBT-IDT: poly(5,6-Dicyano-2,1,3-Benzothiadiazole-alt-Indacenodithiophene)
DC: Direct Current
DSSC: Dye-Sensitized Solar Cell

E

EIS: Electrolyte-Insulator Semiconductor
ETL: Electron Transporting Layer
eV: electron volts

F

FEFS: Field-Effect Ferroelectric Semiconductor

FF: Fill Factor

G

GaAs: Gallium Arsenide
GBHJ: Grade Bulk Heterojunction
Gr: Graphene

H

H: Hydrogen
HHA: Heteroheptacene
HTL: Hole Transport Layer

I

IAS: Improved Amorphous Semiconductor
IB: Intermediate Band
IBSC: Intermediate Band Solar Cell
IMOS: Improved Metal-Oxide Semiconductor
IOS: Inorganic Semiconductor
ITO: Indium Tin Oxide
I-V: Current-Voltage

L

LDS: Luminescent Down-Shifting
LUMO: Lowest Unoccupied Molecular Orbital

M

MBE: Molecular Beam Epitaxy
MCPV: Medium Concentrator Photovoltaic
MicoSem: Microcrystalline Semiconductor
MINDS: Metal-Insulator-Nonuniformly-Doped Semiconductor
MIS: Metal-Insulator Semiconductor
MNN-MIS: Metal-Nanowire-Network-Metal-Insulator Semiconductor
MOIS: Metal-Organic-Insulator Semiconductor
MOLS: Molecular Semiconductor
MOS: Metal-Oxide Semiconductor
MQD-PDES: Multi-type QDs Phot-induced Doping Enhanced Semiconductor
MS: Metal Semiconductor
MS-HDSW: Microstructure Semiconductor based on Heavily-Doped Silicon Wafers

N

nc-Si:H: Hydrogenated Nano Crystalline Si
NFA: Non-Fullerene Acceptor
NFA-BHJ-OSC: Non-Fullerene Acceptor Bulk Heterojunction Organic Solar Cell

NP: Nanoparticle
NS: Nanocrystalline Semiconductor

O

O/I: Organic-Inorganic
O/O: Organic-Organic
OS: Organic Semiconductor

P

PAAM: Porous Anodic Alumina Membrane
Pc's: Phthalocyanines
PAS: Particulate Semiconductor
PCS: Polycrystalline Compound Semiconductor
PCE: Power Conversion Efficiency
PEDOT: poly(3,4-ethylenedioxythiophene)
PS: Polymer Semiconductor
PSS: Polystyrene Sulfonate

Q

QD: Quantum Dot
QD-IBS: QD Intermediate Band Semiconductor
QR: Quantum Ratchet
QW: Quantum Well

S

SB: Schottky Barrier
SCM: Scanning Capacitance Microscopy
SC-NP/O: Semiconductor NP-Organic
SCSSC: Semiconductor-Sensitized Solar Cell
SEM: Scanning Electron Microscope
Si: Silicon
$SiCl_4$: Silicon Tetrachloride
SiH_4: Silane
SIS: Semiconductor-Insulator Semiconductor
SITOS: Sprayed Indium Tin Oxide Semiconductor
SKPFM: Scanning Kelvin Probe Force Microscopy
SN: Sulfur-Nitride

T

TO: Tin Oxide

U

UPS: UV-Photoelectron Spectroscopy

UV: Ultraviolet

V

VB: Valence Band
Voc: open circuit voltage

X

XPS: X-ray Photoelectron Spectroscopy

Semiconductor Materials and Modelling for Solar Cells   Materials Research Forum LLC
Materials Research Foundations **104** (2021)   https://doi.org/10.21741/9781644901434

1.   Introduction

The demand for renewable energy is ever increasing and solar energy is one of them. There are plans to substitute the conventional energy sources based on fossil fuels to photovoltaic generators [1]. In a large-scale photovoltaic conversion system (Figure 1) [2], solar cells are an important component, but this system has other components too embedded in the subsystems such as energy storage elements and power conditioning circuits. In spite of the great reduction in the cost of photovoltaic generation it is required to reduce its cost further by reducing the cost of the solar cells [2, 3]. This may be accomplished by reducing the solar cells active material cost.

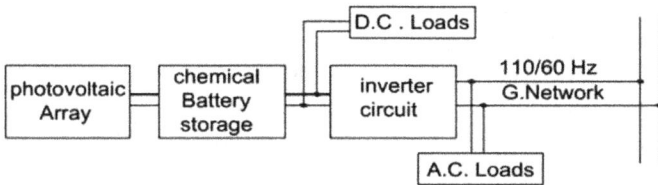

Figure 1   Photovoltaic conversion system [2] (with permission to reproduce/reprint from the publisher).

Solar cells have semiconductor materials which make their active zone. In this zone, photons provide charge carriers which are collected by the solar cells. Another section of the solar cells is a semiconductor body that may be amorphous or polycrystalline or crystalline. The active layer is supplied by two metal electrodes, the cathode and the anode. Then the solar cells are encapsulated in insulating laminates with a double layer of insulating material. The most advanced solar cell technology is the single crystal silicon (Si) solar cells [4]. But the cost of these solar cells is relatively high. Therefore, there is an extensive effort to invent and develop new thin film solar cells [5-40]. These solar cells are made from heterojunctions and their active semiconductor materials are thin film structures.

In this book an almost comprehensive survey about the new and advanced solar cells structures development is presented along with their merits. Such a survey may help researchers to get a comprehensive background for a rapid start in the related research. There is progress in solar cell efficiency because of the improvement in surface recombination velocity, a higher and low charge density, high ultraviolet (UV) sensitivity, etc. [41].

Semiconductor Materials and Modelling for Solar Cells   Materials Research Forum LLC
Materials Research Foundations **104** (2021)   https://doi.org/10.21741/9781644901434

2.   Review of semiconductor technologies in solar cells

The sunlight is one of the most important renewable energy resources to be converted to electricity and produce clean energy. Solar cells are devices for generating the electricity from the sunlight [42]. Today, reducing the solar cell production cost is a key component for manufacturers and customers, and photovoltaic cells are improved based on thin film technology to be used in commercial devices. In fact, solar cells are the most excellent alternative for fossil energy due to their potential inexpensive and eco-friendly nature.

The considerable thin film technologies involve 1) cadmium telluride (CdTe), 2) amorphous Si (a-Si), 3) crystalline Si, 4) copper-based such as copper indium diselenide (CID), copper indium gallium selenide (CIGS), CuO, Cu_xO_y-ZnO:Al, $Cu_2Zn_{1-x}Bi_xSnS_4$(CZBiTS) [43], copper-tin sulfides (CTSs) [44], 5) barium-based such as barium silicide [45], $BiTiO_3$ (BTO) [46] and Bi_2S_3 [47], 6) dye-sensitized TiO_2 and 7) polycrystalline-Si SIS cells [26, 48-55]. In crystalline semiconductors charge carriers (CCs) have much higher mobility by several orders of magnitude than non-crystalline semiconductors [56]. Currently, polycrystalline-Si SIS is one of the highly efficient semiconductor materials in solar cells. These types of solar cells have the benefits because of the hydrogen incorporation and suitable surface mixture which improve the photovoltaic performance in low-quality materials, i.e., amorphous semiconductors [57].

The solar cells are placed in a measuring electrical circuits [58] to measure their I-V characteristics [58]. Figure 2 left depicts typical I-V curves for solar cells. This system of curves shows the effect of the series resistance Rs of the solar cell on its I-V characteristics.

The ideal solar cell has zero R_s. As R_s increases, the I-V curves will be deviate much from its ideal nearly square curve. Figure 2 right [2] shows additionally the effect of the shunt resistance of the solar cell on its I-V curve. As clear from the figure R_{sh} has also a detracting effect on the solar cell. The ideal value of R_{sh} is infinity.

Because the solar cell is merely an illuminated p-n junction diode, it will have the electrical equivalent circuit shown in Figure 3 [2, 42, 59]. The current source I_{ph} represents the generated photo current by incident solar radiation, while I_d is the dark diode current. The series resistance R_s represents the internal series resistance of the solar cell and R_{sh} depicts the internal shunt resistance of the solar cell.

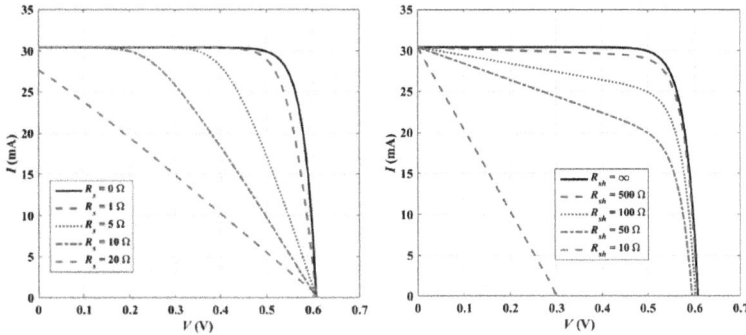

Figure 2   *Schema of I-V curve [2] (with permission to reproduce/reprint from the publisher).*

The basic parameters for determining the I-V characteristic are: the photo current, I_{ph} provided by a solar cell device under illumination at the short circuit current, I_{sc}, the open circuit voltage, Voc, the reverse saturation current, I_s, the ideality factor, n, the series resistance, R_s, and the shunt resistance, R_{sh}.

The illuminated I-V curve follows the model equations [2]:

$$I = I_{ph} - I_s \left(e^{\frac{(V+IR_s)}{nV_T}} - 1 \right) - \frac{V+IR_s}{R_{sh}} \qquad (1)$$

By putting $I = 0$, $R_s = 0$ and $R_{sh} = \infty$, we have,

$$V_{oc} = nV_T \ln\left(\frac{I_{ph}}{I_s} - 1\right) = \frac{kT}{q} \ln\left(\frac{J_{sc}}{I_s}\right). \qquad I_{ph} = J_{sc} \; and \; J_{sc} \gg I_s \qquad (2)$$

Where I_s is the reverse saturation current of an ideal p-n junction, $qV_T = KT$, and q is the electron charge.

The performance parameter of the solar cell is its power conversion efficiency PCE defined by the ratio of the maximum power output, $P_M = (V_{Max} I_{Max})$ and the input incident solar radiation power Pin. (Eq. 3). In fact, solar cell PCE, which is depicted by η, depends on J_{sc} and V_{oc} [2]. It follows that:

$$\eta = \frac{V_{Max}I_{Max}}{P_{in}} = \frac{FFV_{oc}I_{sc}}{P_{in}} \qquad (3)$$

P_{in} for AM1=$100mW/cm^2$, and FF is the Fill Factor which measures the squareness of the illuminated I-V curve and is given by equation 4 [2],

$$FF = \frac{V_{Max}I_{Max}}{V_{oc}J_{sc}} \tag{4}$$

The PCE depends largely on the energy gap of the semiconductor material of the solar cell getting its peak at a bandgap of about 1.35 electron volt (eV) [52].

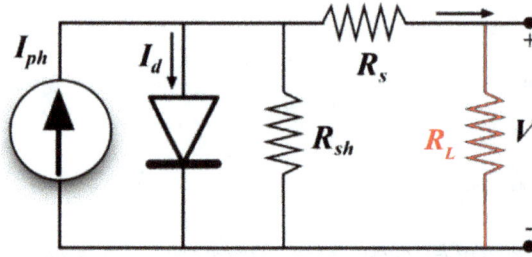

Figure 3   *Equivalent circuit of solar cells.*

The solar cell is operated at the maximum power point to extract the highest possible electrical power from it Pmax. At this point the voltage of the cell is Vmax and the current is Imax. To operate the solar cell at Pmax one loads it by a resistive load RL as shown in Figure 3. And given by Eq. 5 [60],

$$R_L = \frac{Vmax}{Imax} \tag{5}$$

Any solar cell operates in a way that photons of the sun create a photocurrent after hitting the solar cell (Figure 4) [2], and pump electrons from the valence band (VB) to the conduction band (CB). The semiconductors have VBs filled by electrons and an empty conduction band which two bands are isolated from each other by an energy gap, E_g of about a few electron volts, e.g. for Si it is 1.1eV and for gallium arsenide (GaAs) it is 1.45eV, where each atom is joined to the adjacent atoms by covalent bonds (Figure 5) [2]. These atoms are fixed and do not have any conducting electricity, so doing some work – i.e., heating, doping, illuminating, etc. causes their bonds to break. Afterwards, E_g separates the free electrons from the free holes and the electrons move towards the CB (Figure 5) [2].

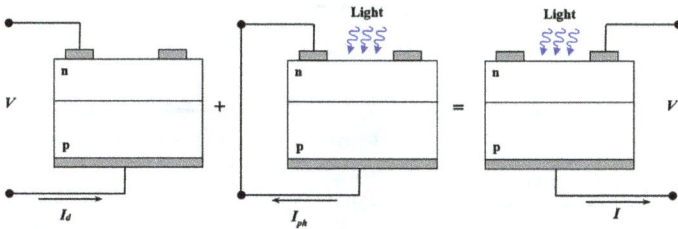

Figure 4   Photon emission [2] (with permission to reproduce/reprint from the publisher).

This happens when the photon energy is high enough (higher than or equal to Eg). If the photon energy is higher than the E_g, the excess energy is converted into heat which is not helpful to provide power. Also, photons with an energy less than the E_g do not have enough energy for the electrons to reach the CB, so falling again to VB [2, 61].

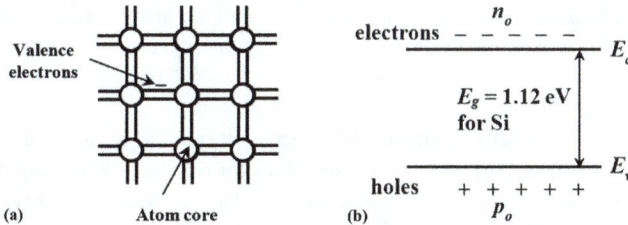

Figure 5   (a) VB and covalent bands, (b) separation of electrons and holes [2] (with permission to reproduce/reprint from the publisher).

In this book, the semiconductor materials in solar cells are reviewed where researchers have tried to solve important problems, such as lattice match, energy band alignments, recombination of minority carriers in bulk and at the interface, interface chemical stability, for solar cells to enhance their performance, thin film quality and scalability [48].

3.   Semiconductor materials in solar cells

The main aim of manufacturing other new semiconductor materials for solar cells has been to improve the ability of the solar cells to harvest the solar radiation [62-65]. Figure 6 [64] depicts the most important light harvesting semiconductor-based systems for solar cells [64, 65].

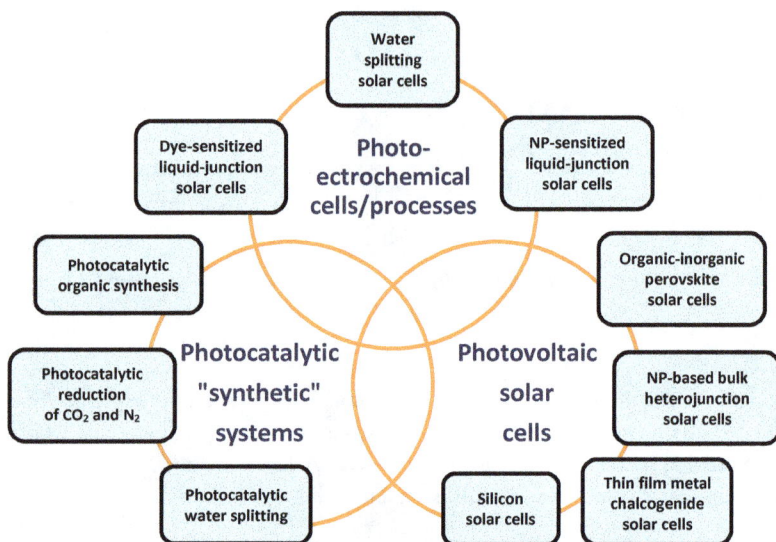

Figure 6   *The most important light harvesting semiconductor-based systems for solar cells [64] (with permission to reproduce/reprint from the publisher).*

Currently, most of the solar cell markets have been allocated to crystal and polycrystalline semiconductor materials [66], but other semiconductor materials also have their markets. Figure 7 [64] shows the evolution of semiconductor solar cells from the first generation using single and polycrystalline Si solar cells to the third generation using nanocrystalline semiconductors through the second generation thin film p/n junction solar cells. This section describes and reviews the semiconductor materials in solar cells.

The following sections will survey the different advanced solar cell structures.

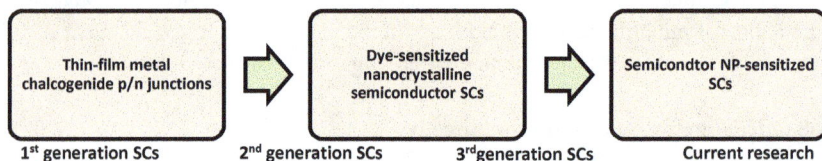

Figure 7   *The evolution and progress of semiconductor solar cells [64] (with permission to reproduce/reprint from the publisher).*

3.1   Organic semiconductor (OS)

The class of the Organic Semiconductor (OS) solar cells received intensive research leading to an appreciable increase of their optical conversion efficiency. OSs solar cells in their simplest form are made with the organic film embedded between two thin metallic electrodes which are optically transparent (Figure 8) [67, 68]. They are built on different substrates. They are characterized by light weight and less cost. They have generally good optical performance but they possess relatively bad electrical performance.

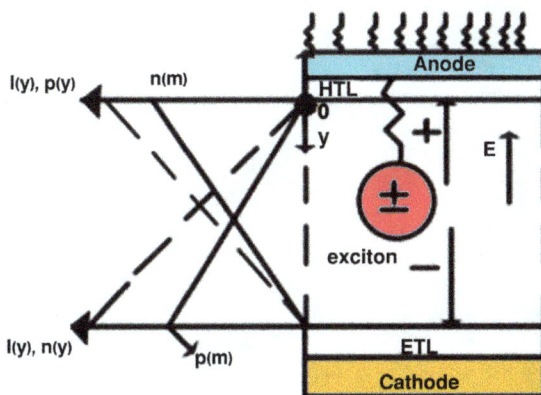

Figure 8   Configuration of OS [68]: n(m) and p(m) are maximum electron and hole density, respectively; n(y) and p(y) are electron and hole concentration, respectively density (with permission to reproduce/reprint from the publisher).

In Figure 8, the electron current is distributed reversely when it is zero at the HTL and maximum at the ETL. So, the hole distribution tracks the current of the hole and the electron distribution tracks that of the electron.

Figure 9 shows the chemical structures of a few organic semiconductors utilized as an operative material in solar cells [69]. They include CuPc (fluorinated), MEH-PPV, P3HT, $PC_{60}BM$, $PC_{70}BM$, PCDTBT, poly(3,4-ethylenedioxythiophene) (PEDOT): Polystyrene Sulfonate (PSS), PMMA, PTB_7, and spiro-OMcTAD, etc. [70].

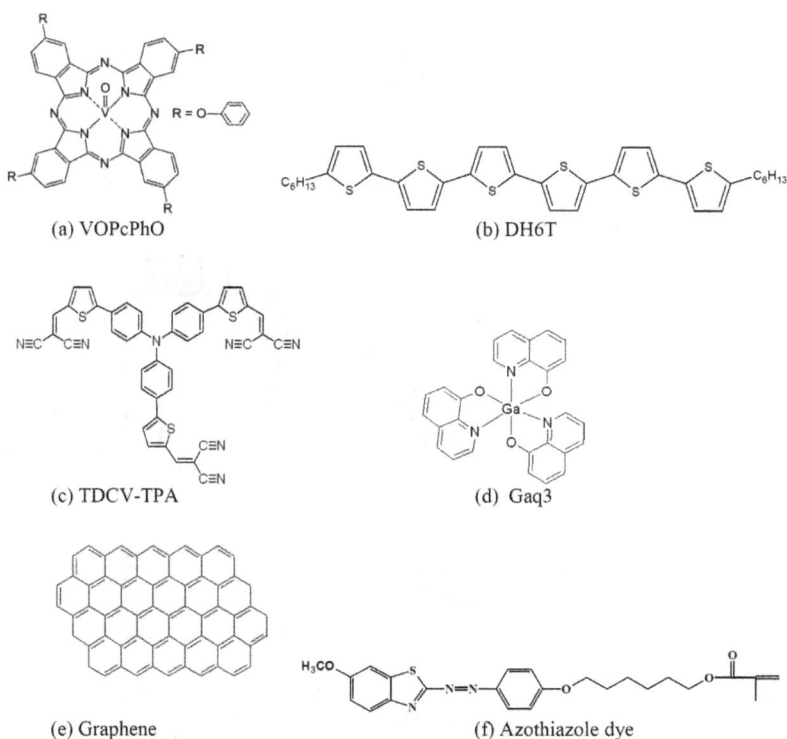

Figure 9   *Chemical structures of some organic semiconductors [69] (with permission to reproduce/reprint from the publisher).*

As one sees the OSs are molecular in nature in opposite to the metallic semiconductors which are atomic in nature. So, one finds great difference between the properties of the organic and metallic semiconductors. Normally the organic semiconductors have lower dielectric constant, high optical absorption coefficient and low mobility in addition to relatively high exciton dissociation energy [68]. They are also called Molecular Semiconductors (MOLS) [71]. The design of OS is empirical, because by discovering the materials and modifying their structure and adding low chemicals, it is possible to control the layer texture deposition. The majority of prior work on these kinds of semiconductors have been done by single-dye Schottky barrier (SB) solar cells, but p-n heterojunction solar cells, e.g., $SnO_2/CuPc/BP/Ag$,

SnO_2/BP/CuPc/Ag and $Zn_xCd_{1-x}S$/ZnTe [72, 73], etc. are better than SB solar cells. The p-n heterojunction cells – i.e., the organic dyes with contacting metal electrodes or oxide (organic-inorganic (O/I) heterojunctions), heterojunctions formed between different organic dyes (organic-organic (O/O) heterojunctions), and semiconductor Nanoparticles (NPs) with organic host ligands and polymers (semiconductor NP-organic (SC-NP/O) heterojunctions)– have the dye layers as pairs with 300Å thick which absorb 70% of the solar radiation. Currently, non-fullerene acceptor bulk heterojunction organic solar cells (NFA-BHJ-OSCs) have also attracted interest of markets due to their high PCE and stability in comparison with semiconductor solar cells such as fullerene acceptor solar cells [74]. Materials for OS layers usually are perylenes, phthalocyanines (Pc's), perylene-diimides-based small molecules [75], ladder-type heteroheptacene (HHA) based on small molecules [76] which have high Fermi-level separation, stable temperature and high absorption and some manufacturing techniques such as X-ray photoelectron spectroscopy (XPS) and UV-photoelectron spectroscopy (UPS) are used for their characterization. In p-type heterojunction cells, the electrons move to the Ag layer and holes move to the SnO_2 layer, while in n-type heterojunction cells the reaction is reversed [72, 77, 78]. In single-dye SB, after a period of time, they lose their light due to metal contact and their light are provided by absorption spectrum. This causes their work to be limited. Thus, this limitation must be set by energy transfer, called Förster, from single-dye layer to the metal contact which produces heat transfer instead of carriers [77]. Figure 10 [77] shows the structures of single-dye SB solar cells and p-n heterojunction solar cells with two thin organic layers on Indium Tin Oxide (ITO)/Tin Oxide (TO).

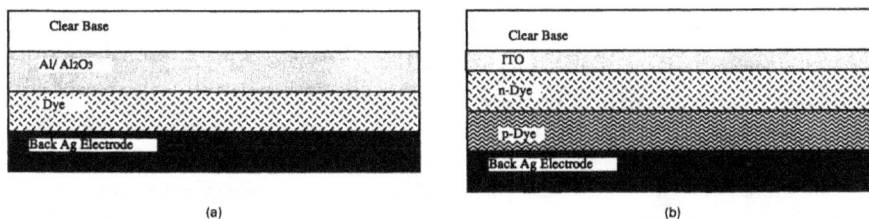

Figure 10   The structures of (a) single-dye SB solar cells and (b) p-n heterojunction solar cells [77] (with permission to reproduce/reprint from the publisher).

Today attempts for creating solar cells according to heterojunction organic polymers and dyes, or combination of these materials with different inorganic NPs, etc. continue to enhance the processing parameters to inhibit light loss which plays an important role in new technologies. ITO can prevent light loss, because it does not require any set by energy transfer. Also, deposition of ITO on other organic and inorganic semiconductors such as CdS can promote their electrical and crystallization [77-79]. Researchers have shown that aluminum zinc oxide (AZO) as a cheap oxide can be used for several applications, except when there are moisture issues, without any worrying about the performance of the cell compared to ITO. It is less costly as well as has good optical performance in solar cells [80].

3.2   Polymer semiconductor (PS)

Polymer Semiconductor (PS) is a type of OSs. It is a solar cell whose junction is made by a polymer interface such as polymer-fullerene with PCE of 8.13% [42], polymeric Sulfur-Nitride (SN) placed on GaAs with 20.7 volts Voc, Metal-GaAs with voltage 40% more, and polymer-fullerene-based bulk heterojunction with a great stability as well as high electrical conductance [81]. A simple vapor deposition technique is employed for PS solar cells which does not need any oxidation procedures used in Metal-Insulator Semiconductor (MIS) structures which have the same Voc [82-86]. Today, n-type PSs such as poly(5,6-Dicyano-2,1,3-Benzothiadiazole-alt-Indacenodithiophene) (DCNBT-IDT), are used as a narrow band gap polymer with much higher absorption coefficient and remarkable PCE of 8.32% with a small energy loss [87]. Also, recently, graded bulk heterojunction (GBHJ) PSs due to having an active layer of donor-blend-acceptor structure, has paid much attention in developing solar cells. They have the PCE of 10.15%. But their PCE is enhanced by using nonfullerene acceptors (NFAs) and reaches 12.49% with J_{sc} of 19.88 mA/cm^2, Voc of 0.798 V, and FF of 78.68% [88].

3.3   Inorganic semiconductor (IOS)

Inorganic semiconductors (IOSs) are known due to their low cost and high efficiency [89]. They are classified in five groups: 1) group-IV (Si, Ge, hydrogenated amorphous Si (a-Si:H), etc.), 2) III–V (GaAs, GaP, InP, InAs, AlAs, etc.), 3) II–VI (CdTe, CdS, ZnTe, ZnSe, ZnS, etc.), 4) I–III–VI$_2$ (CuInSe$_2$, CuGaSe$_2$, etc.) and 5) I$_2$–II–IV–VI$_4$ (Cu$_2$ZnSnSe$_4$, Cu$_2$ZnSnS$_4$, Cu$_2$ZnGeSe$_4$, etc.) semiconductors [67, 90]. Today, using IOSs such as inorganic NPs which are made more by nanocrystal assembly method [91] and their bandgap increases by decrease of the nanocrystal diameters, particularly those made from II-VI semiconductors (SC-NP), e.g. CdSe (Table 1) [42], porous anodic alumina membrane (PAAM) such asCuInSe$_2$, CdS, CdTe,Bi$_2$Te$_3$, Sb$_2$S$_3$, AgSbS$_2$, andmain metaloxides such as V$_2$O$_5$,ZnO [61, 79, 92-97] and TiO$_2$, can be used for polymer-inorganic solar cells which PCE on polymer bandgap and the lowest unoccupied molecular orbital (LUMO) depends on inorganic acceptors, and have increased in new solar cell technologies due to electron transporting and light harvesting [42, 78, 98-108]. The solar cells which use inorganic NPS are called semiconductor-sensitized solar cells (SCSSCs), e.g. BiVO$_4$ [45, 97, 109, 110]. IOSs have many applications especially in PSs. They altogether with PSs can create polymer-IOS with PCE exceeding 3% due to harvesting more sunlight, but their PCE is lower than other PSs such as polymer-fullerene solar cells. There are three types of polymer-IOSs: a) bulk heterojunctions, b) bilayer heterojunctions and c) ordered heterojunctions. Polymer-IOSs are very cost-effective. The geometry of them – i.e., interface between IOS and polymers– have more effects on performance of solar cells [42, 101, 102]. Currently, the IOS solar cells are fabricated by processes such as screen printing, ink jet printing, and dip coating, roll-to-roll processing and painting [42, 111, 112]. Figure 11 shows the polymer-inorganic and inorganic NPs semiconductors [42].

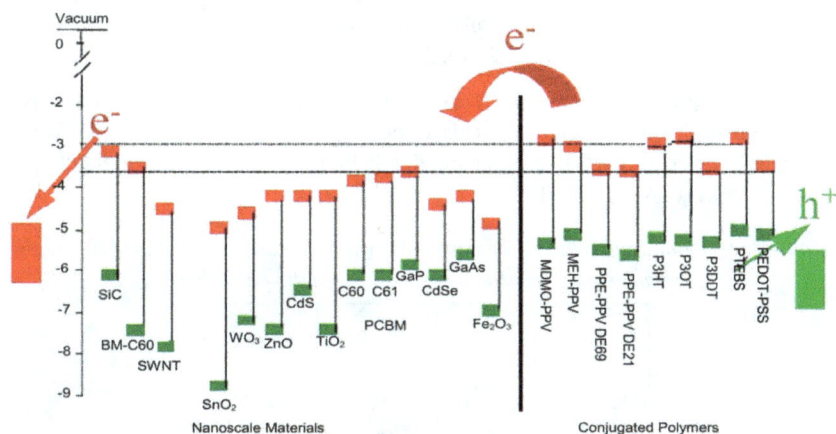

Figure 11   Polymer-inorganic and inorganic NP semiconductors utilized in solar cells [42] (with permission to reproduce/reprint from the publisher).

Table 1   Different CdSe used for polymer–inorganic hybrid solar cells (D: diameter) [42] (with permission to reproduce/reprint from the publisher).

Name	Size (nm)	Polymer	η (%)
CdSe nanocrystals	5 nm (D)	MEF-PPV	0.2
CdSe nanorods	8(D) × 13 nm	P3HT	0.2
CdSe nanorods	7(D) × 60 nm	P3HT	1.7
CdSe nanorods	5(D) × 65 nm	OC_1C_{10}-PPV	1.7
CdSe nanorods	5(D) × 65 nm	P3HT	2.6
CdSe nanorods	15(D) × (30~100) nm	P3HT with additives	2.65

CdSe tetrapods	5(D) × 50 nm	OC_1C_{10}-PPV	1.8
CdSe tetrapods	5(D) × 50 nm	OC_1C_{10}-PPV	2.8
CdSe tetrapods	5(D) × (30-50) nm	PCPDTBT	3.13
CdSe multipods	5(D) × (30~70) nm	APFO-3	2.4
CdSe hyperbranched nanocrystals	150 nm	P3HT	2.18
CdSe quantum dots	5.5 nm (D)	P3HT	2

The low bandgap polymers are very promising for more increase of polymer-inorganic hybrid solar cell efficiency [42, 89]. Table 2 shows the bandgaps and E_g of IOSs [42].

Today, a technique as known Atomic Force Microscopy (AFM) is very effective for determination of IOSs characteristic. This technique involves three stages: 1) Scanning Capacitance Microscopy (SCM), 2) Scanning Kelvin Probe Force Microscopy (SKPFM), and Conductive AFM (C-AFM) [113].

Table 2   *Bandgaps and E_g of some IOSs with crystalline structure [42] (with permission to reproduce/reprint from the publisher).*

Name	E_g(eV)	VB (eV)	CB(eV)
CdSe	2.1	-5.8	-3.7
TiO_2	3.2	-7.4	-4.2
ZnO	3.5	-7.9	-4.4
PbSe	0.8	-5.0	-4.2
PbS	0.4	-.371	-3.3
CdTe	2.4	-6.3	-3.9
CdS	2.42	-6.92	-4.5
Si	1.12	-5.12	-4.0
SnO_2	3.6	-7.9	-4.3
$CuInS_2$	1.5	-6.0~-5.0	-3.7~-4.1
$CuInSe_2$	1.0	-5.6	-4.6

3.4   Particulate semiconductor (PAS)

Particulate semiconductors (PASs) are fabricated by a new method called high-speed emulsion coating which they are deposited in a binder. For example, one of the particulate semiconductors is particulate ZnO which has 3% PCE, and above 50% quantum efficiency with UV light. They are employed as photoconductive layers for photocopying applications which require electrofax layers and have only 1/4 μm diameter with a thickness of 10 μm in contrast to many monolayers. The high-speed emulsion coating is a cost-efficiency method using a paper substrate for preparing the electronic materials and it does not change the chemical state of the particulate semiconductor comparing with other techniques such as chemical spray or vacuum deposition. Totally, coating technology can create large areas in comparison to other techniques [114].

3.5   Conductor polymer semiconductor (CPS)

In conductor PS (CPS), the polymer film is sandwiched in a conductor [115]. In this semiconductor, the top layer can be a SB [115-119], oxide, organic or inorganic, as known heterojunctions, [120-122], electrolyte [123-125], or mixed of them [126]. Commonly, the characteristic of this semiconductor is linear [116-119] but its efficiency is poor due to its weak FF and J_{sc} because of recombination of free electrons and holes on the interface states. The novel CPSs are PEDOT family which they have a better PCE but they do not have the strong FF and J_{sc} [2]. Typical best junctions are given in Table 3 [126].

Table 3   Best different types of the CPSs [126] (with permission to reproduce/reprint from the publisher).

Name	Voc (mV)	FF	η (%)
$Al\|(CH)_x\|Au$	400	0.25	0.2
$Al\|(CH)_x\|E + 5O_2$	310	0.21	0.1
$Al\|(CH)_x\|E + 5O_2$	650	0.30	0.1
$Al\|(CH)_x\|E + 5O_2$	246	0.15	-
$In\|(CH)_x\|E + 5O_2$	100	0.25	0.025
$Pb\|(CH)_x\|E + 5O_2$	173	0.25	0.002
$Pb\|(CH)_x\|E + 5O_2$	103	0.25	6.3×10^{-5}
$n - CdS/p - (CH)_{yx}$	300	0.4	0.5
$n - CdS/p - (CH)_x$	300	-	-

$n - Si/p - (CH)_x$	530	0.32	4.3
$n - Si/PP_y$	300	0.46	1.24
$Na_2S + S_6|(CH)_x$	300	-	-
$In|(PcGeO)_n|SnO_2$	100	-	5×10^{-4}
$SnO_2|2PVPI_2|Pt$	65	0.27	0.09

Also, this solar cell is highly dependent on the polymer sample and the way that contacts are prepared. Nevertheless, according to Figure 12 which shows the evolutions from 1970 to 1980 and after, the PCE of such cells is still not high [126, 127].

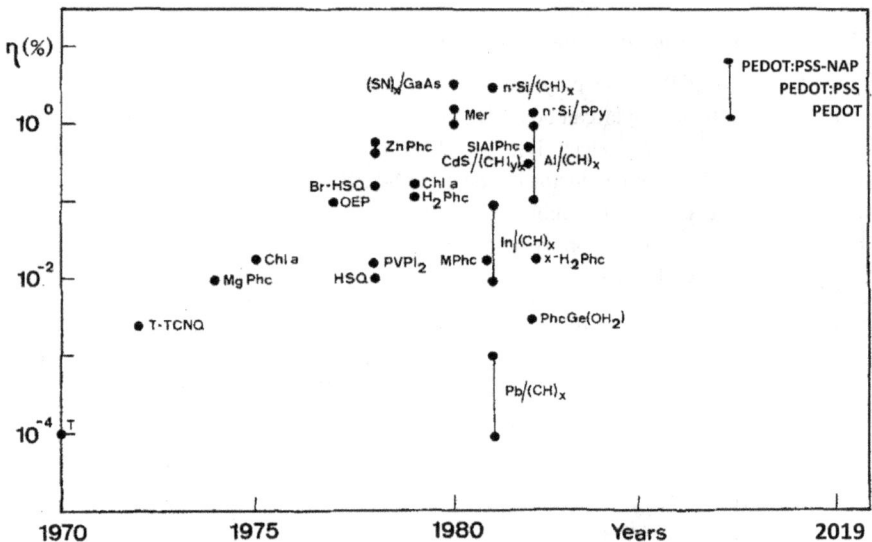

Figure 12   *Evolution of η(%) against years [126, 127] (with permission to reproduce/reprint from the publisher).*

3.6   Metal semiconductor (MS)/Schottky barrier (SB)

In some terrestrial and space applications, it is required to convert solar energy to electrical energy. So, metal semiconductor (MS)/SB solar cells have a potential for them. The cell consists of the semiconductor material supplied with a thin transparent metal film to allow the passage of light while building a potential barrier with the substrate material. This thin metal layer will modify the refractive index, n_{AR}, at the surface of the substrate since it has a n_{AR}. Because of a relatively high minority-carrier collection efficiency, SB solar cells have a good current output especially when the E_g is direct –i.e., GaAs. So, for using this helpful characteristic, the sunlight amount coupled into the solar cells has to be increased by a proper antireflection (AR) coating as depicted in Figure 13 [128]. n_{AR} and the thickness, t_{AR}, of such antireflective coating can be expressed by the relations (Eqs. 6,7) [128, 129].

$$n_{AR} = (n_0 \frac{V^2}{(U-n_0)} + U)^{1/2} \tag{6}$$

$$t_{AR} = \frac{\lambda}{2\pi n_{AR}} tan^{-1} \left(\frac{n_{AR}(U-n_0)}{n_0 V} \right) \tag{7}$$

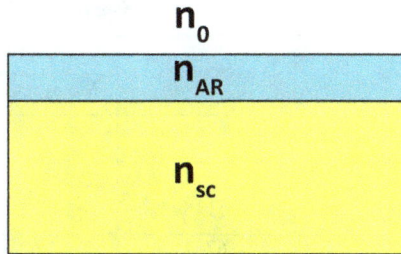

Figure 13   AR coating improving for the MS solar cell [128] (with permission to reproduce/reprint from the publisher).

Where U and V are the real and the imaginary parts of the complex n_{AR}, respectively, n_0 is the refraction coefficient of the air, and n_{SC} is the refraction coefficient of the coated material while λ is the wavelength of the incident. The most suitable coating is that makes the sunlight maximum [128, 129]. Table 4 and 5 shows the data for AR-coated GaAs and GaAs$_{0.78}$P$_{0.22}$SB solar cells, respectively [129].

Table 4   Some data for AR-coated GaAs Sb solar cells [129] (with permission to reproduce/reprint from the publisher).

Sample[a]	t_{AR}(Å)	n_{AR}	Wavelength (Å)	Reflectance, %	
				Measured	Calculated
1	591	2.10	6764	1.35	1.15
			6471	0.72	0.68
			5682	2.2	2.17
			5145	7.2	6.65

			6764	2.04	1.87
2	690	2.12	6471	3.55	3.38
			5682	11.5	11.1
			5145	20.8	19.3
[a]60-Å gold on GaAs coated with Ta_2O_5 evaporated by resistance-heated tungsten boat at 4 Å/s.					

Table 5   *Some data for AR-coated GaAs0.78P0.22Sb solar cells [129] (with permission to reproduce/reprint from the publisher).*

Sample[a]	t_{AR}(Å)	n_{AR}	Wavelength (Å)	Reflectance, %	
				Measured	Calculated
1	550	1.98	6764	6.21	6.18
			6471	4.74	4.45
			5145	1.62	2.19
2	612	2.01	6764	2.12	2.91
			6471	1.46	1.82
			5145	5.78	5.83
3	582	2.03	6764	3.14	4.31
			6471	2.09	2.60
			5145	3.54	3.67
[a]60-Å gold on GaAs coated with Ta_2O_5 evaporated by e-beam gun at 4 Å/s.					

SB solar cells have some benefits including 1) flexibility with polycrystalline thin films as well as direct bandgap semiconductors, 2) less temperature need without any degradation and less cost barrier production, 3) strong radiation resistance, and 4) strong current outputs [86, 130].

3.7   Metal-oxide semiconductor (MOS)

Metal-oxide-semiconductor (MOS) solar cells are made by the growth of a very thin oxide interfacial layer on the metal semiconductor. This oxide layer increases their Voc and efficiency by over 60% compared to solar cells with no oxide layer. For growing the oxide layer, some oxidation methods such as XPS, plasma oxidation, thermal oxidation, and anodic oxidation [2] are used which produce the chemical structure and necessary oxide thickness. These techniques can use water-vapor saturated oxygen, ozone, oxygen gas discharges and other methods such as apparent crystallographic orientation impacts. MOS requires the compounds such as ternary or GaAs due to their advanced growth methods which use vapor-/liquid-phase epitaxy. In fact, the high light absorption in these compounds is good for the use of thin films because of the barrier location at the semiconductor surface. MOS solar cells can be used instead of p-n junction solar cells because of their low cost [86, 131].

Many research [132-135] illustrates that the appearance of a thin oxide layer (10-20Å) between the semiconductor and the metal improves Voc of a SB solar cells without any decrease in its J_{sc}. But using Ag, e.g., ultrathin Ag films, as a barrier leads to disruption in interconnections which results in the decrease of FF, J_{sc}, and also Voc, so the initial oxide must not be thicker than the optimal one [131, 136]. Figure 14 shows the Current-Voltage (I-V) properties of a MOS solar cell for different oxide layer thickness in Å, δ [131].

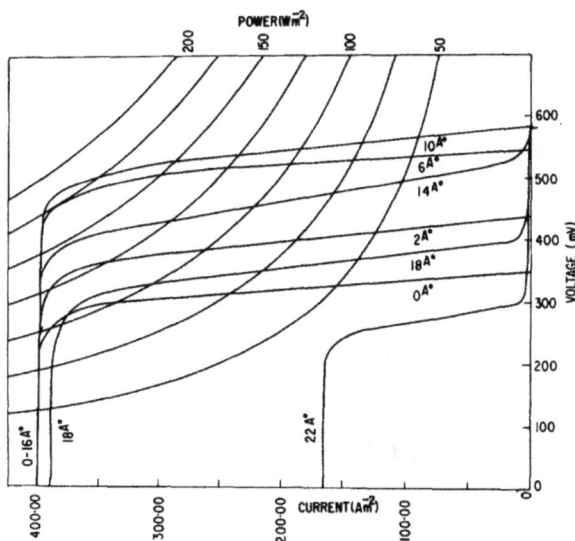

Figure 14   *A cell I-V characteristic of n-type Si base layer (Resistivity = 1 ohm.cm) for different thicknesses, δ, with short current, $J_{sc} = 400 Am^{-2}$ and hole current, $J_p = 10^{-8} Am^{-2}$ at $\delta = 0$ [131] (with permission to reproduce/reprint from the publisher).*

The light intensity, J_{sco} is taken to be 400 Am^{-2} and surface states to have a uniform distribution up to an energy range of 0.3 in eV near the gap center. The barrier height for the surface states, χ_s, barrier height for electrons, χ_n, and barrier height for holes, χ_p are assumed to be 3.8, 3.25, and 3.65eV, respectively [131].

Totally, in MOS, effective mass, m^*, and E_g play a significant role in choosing solar cell characteristics [137, 138]. Also, when the barrier height decreases, the image force effects have an important role in determination of their I-V characteristic, e.g., for a MOS such as Au-SiO₂-nSi MOS, the image force 11% decreases the Voc which cannot be removed [139].

3.8   Improved metal-oxide semiconductor (IMOS)

Improved MOS semiconductor solar cells are a thin layer, e.g., $3\mu m$ n-type GaAs film grown on n+ wafers, e.g., semi-insulating, by liquid phase epitaxy in the (110) orientation. Their PCE in over 60% and also Voc and J_{sc} are high. The thickness of the film is determined by stain and cleave method and resistivity and the Hall coefficient by Van der Pauw's measurements. Then, HCI is used to dissolve any oxide layer and afterwards replaced with methanol and dried with nitrogen. Afterwards, 3000Å of the mixture is evaporated in gas form, i.e., 85% N_2 and 15% H_2, at 450°C for only a few minutes. Then the sample is divided into 2 pieces by one of the pieces having Ga₂O₃, and the concentration is controlled by dilution – i.e., methyl alcohol (Figure 15). Figure 16 shows the characteristic of the IMOS with Voc = 0.55V, J_{sc} = 17.1 mA, and FF = 0.66 [4].

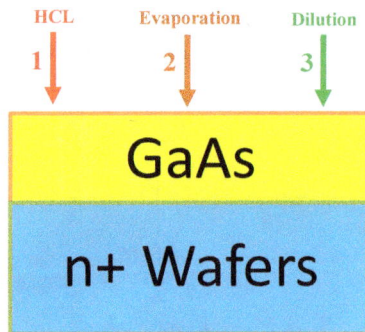

Figure 15   IMOS creation method.

Another technique for creating IMOS is the use of deposited Bi₂O₃ interfacial layer using the methods such as boat thermal evaporation and electron beam evaporation. This semiconductor also shows better Voc, but ones with irregular surface are 3% lower than the flat surface ones, whereas J_{sc} of the irregular surface ones is 30% higher than the flat surface ones. These considerations show that the Voc is dependent on pinhole density, and overall, the PCE of the irregular surface ones is better with efficiency of 9-10% and FF of 76-79% [140, 141].

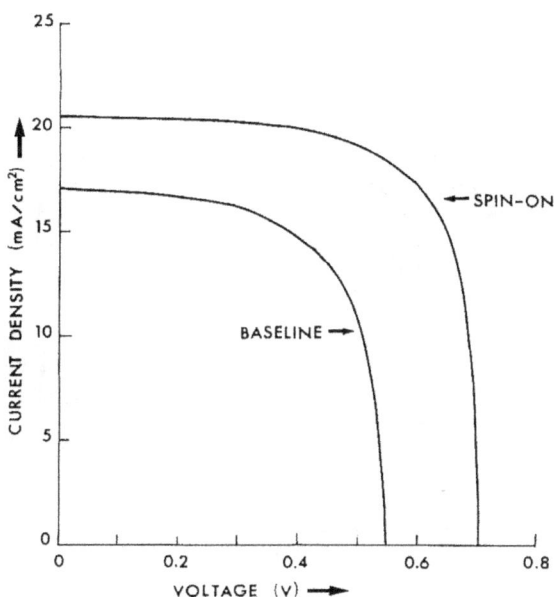

Figure 16   I-V curves for IMOS solar cell [4] (with permission to reproduce/reprint from the publisher).

3.9   Antireflection metal-oxide semiconductor (AMOS)

AR metal-oxide-semiconductor (AMOS) solar cells are made by an additional low temperature process for the growth of an interfacial oxide layer on GaAs which increases PCE and Voc over 60%, and the oxides used are composed of Ga_2O_3 which is very changeable according to its crystalline shape. But it is necessary to know that GaAs cannot oxidize to a proper thickness with dry oxygen alone. Figure 17 [86] shows some light I-V curves for a

(100) surface oxidized by water-vapor-saturated O_2 at room temperature for 114 h and ozone-treated (100), (111), and $(\overline{111})$ GaAs (120°C, 20 min) [86, 132].

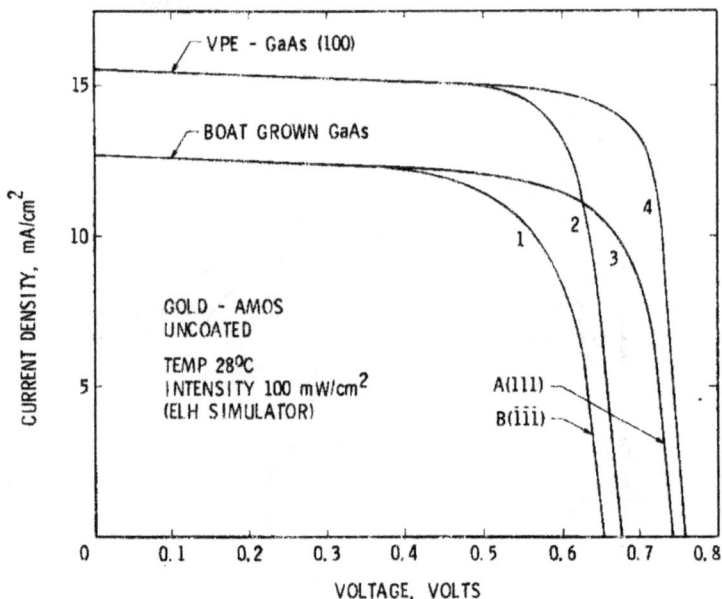

Figure 17   The light I-V properties of AMOS solar cells with O_3 oxide (curves 1, 2, and 3) and O_2-water vapor (curve 4) [86] (with permission to reproduce/reprint from the publisher).

3.10   Sprayed indium oxide (ITO) semiconductor (SITOS)

One tried to substitute the very thin metal layer by a transparent conductive oxide such as ITO. Sprayed ITO semiconductor solar cells (SITOS) (Figure 18 [142]) are provided by mixing of indium chloride and tin dissolved in methanol. This mixture is sprayed on the base layer under steady nitrogen flux and the equations of chemical reactions are as follow (Eqs. 8,9) [143],

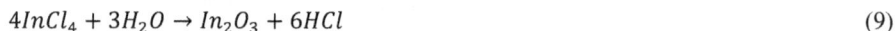

$$SnCl_4 + 2H_2O \rightarrow SnO_2 + 4HCl \tag{8}$$

$$4InCl_4 + 3H_2O \rightarrow In_2O_3 + 6HCl \tag{9}$$

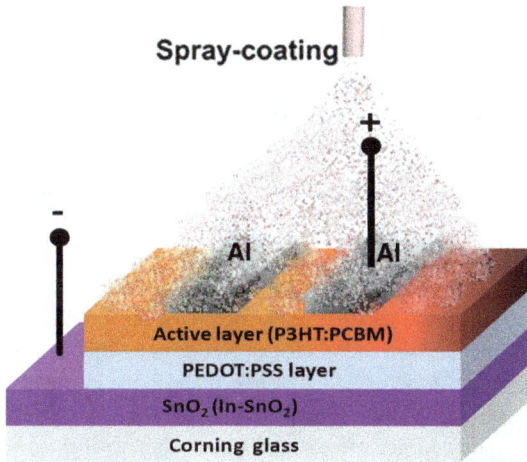

Figure 18 Scheme of SITOS [142] (with permission to reproduce/reprint from the publisher).

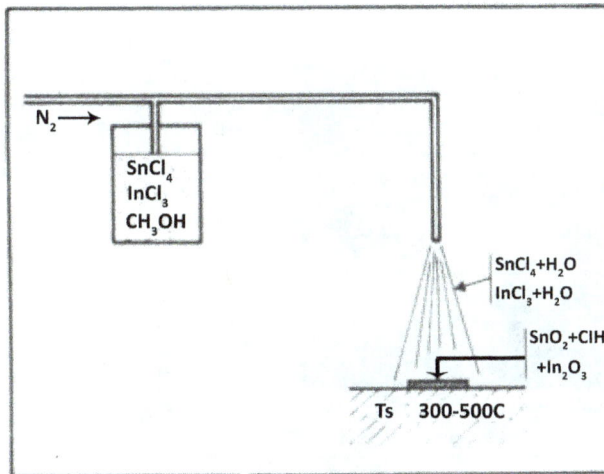

Figure 19   This is the schematic of the apparatus for the deposition of ITO [143] (with permission to reproduce/reprint from the publisher).

The major parameters of the reactions are: 1) the substrate temperature, T_s, which is > 300°C and < 500°C, 2) the atomic ratio, $C_{Sn} = Sn/In$ in the solution, 3) the growth rate ($\sim 0.15 \mu m/min$), 4) the layer thickness around $0.4 \mu m$, 5) the mean transparency is about 90%, and 6) the sheet resistance of the best ITO, e.g. $10\Omega/square$ for $0.4 \mu m$ layer thickness. Figure 19 [143] shows the steps for fabrication of sprayed ITO.

The PCE of this solar cell is 0.56-1.06% for organic solar cell applications [144].

3.11   Conductor-insulator semiconductor (CIS)

This type of solar cells is composed of a base or substrate layer. The top layer is the conductor layer while the insulator layer is sandwiched between the upper and the lower layer. Figure 20 [3] depicts the schematic construction of such type of solar cells. The insulator layer plays an important role in the performance of the solar cell and it can be a thin polymer film, oxide semiconductor or sensitizing dye on ultrathin, called ultrathin insulating layer, which acts as a buffer, decreases the image force effects, and is very thin, to allow CC tunneling across it with little effect on the current as J_{sc} in the devices. Other current transport by hopping via trapping or pinholes in insulator is possible, but these methods have secondary importance [3, 145]. For optimization of the CIS device, the conductor work function and the thickness of the interfacial layer, e.g., an oxide of the base semiconductor, are the necessary parameters for the selection of the top contact [3, 146-149].

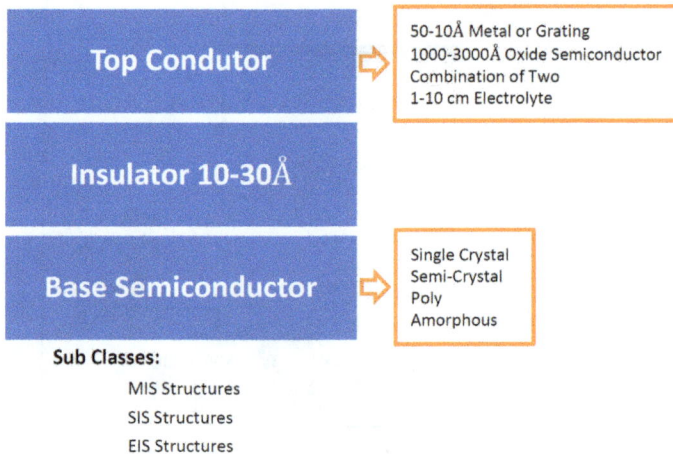

Figure 20   *Schematic of CIS solar cells [3] (with permission to reproduce/reprint from the publisher).*

For p-type base semiconductors, the work function of the top layer should be low, i.e., less than or equal to the electron affinity and the band gap of the base semiconductor, but for n-type base semiconductors, the top layer work function must be high – i.e., greater than or equal to the sum of the electron affinity and the band gap of the base semiconductor [3]. Thus, one of the major benefits of CIS semiconductors is that they can be used as a p-n junction diode in electronic appliances [134, 146, 150, 151], because the conductivity of the base semiconductor can be changed by true selection of the base semiconductor and the top layer [3]. So, they can also be used in large-scale photovoltaic solar energy conversion systems [152]. They have low cost and their operating lifetime is about 20 years [3].

The major parameters that have an effect on the CIS solar cell performance are as follows [3]:

1) Top layer work function,

2) Interfacial layer thickness,

3) Base semiconductor parameters such as diffusion length, mobility and minority carrier lifetime variation, Fermi level, intrinsic carrier concentration and absorption coefficient, which all of them rely on the doping level.

4) Surface states and surface charge,

5) Insulator charge,

6) Temperature,

7) Incident solar radiation intensity

8) Minority carrier lifetime

9) Crystallographic orientation, and

10) Base semiconductor band gap.

3.11.1 Metal-insulator semiconductor (MIS)

A subclass of the CIS cells is the MIS solar cells where one uses an interfacial insulating layer, i.e., oxide, sulphide or combination of them between the metal, and the semiconductor which can highly increase the photovoltaic PCE [130, 132, 147, 153-159] and the top layer of that is a metallic compound, e.g. (SN)x [3]. An E_g diagram for this structure has been given in Figure 21 [130] in which a p-type semiconductor as well as the biased semiconductor with top metal joint by voltage V_a have been shown. The band gaps for the semiconductor and insulator are E_{gs} and E_{gi}, respectively. The energy difference between the insulator and semiconductor band edges – i.e., 3.2 eV for Si-Si dioxide – is Φ_{si}. Also, the metal-to-insulator barrier height is Φ_{mi} [130]. Some metals for p-type MIS solar cells are chromium, titanium, aluminum, beryllium, magnesium, tantalum, etc. Also, some metals for n-type MIS solar cells are palladium, silver, platinum, gold, CdTe, etc. It is very important to know, in p-type solar cells, all parameters reduce with gamma radiation exposure (Figure 22 [160]), so for example for AR coating [161], the spinning technique must be done carefully.

Up to now, two methods have been used for MIS fabrication: 1) grating method, 2) transparent metal method, the first method is used more, and four methods are also used for improvement of PCE: 1) reduction of the surface recombination velocity, 2) design of a high-low doping profile at the back of the solar cells, 3) semiconductor doping, and 4) optimization of the thickness [3, 160, 162, 163]. Figure 23 [3] depicts two structures of MIS solar cells.

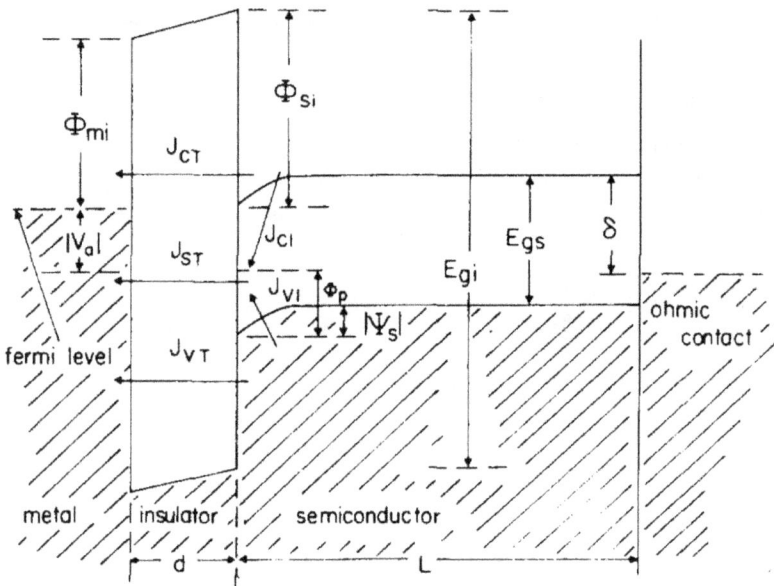

Figure 21   Scheme of E_g diagram of the MIS solar cell with a p-type semiconductor. The current flows from the VB and CB of the semiconductor to the metal joint are J_{VT} and J_{CT}. Also, J_{ST} is a flow from surface states to the metal supplied by J_{VI} and J_{CI} [130] (with permission to reproduce/reprint from the publisher).

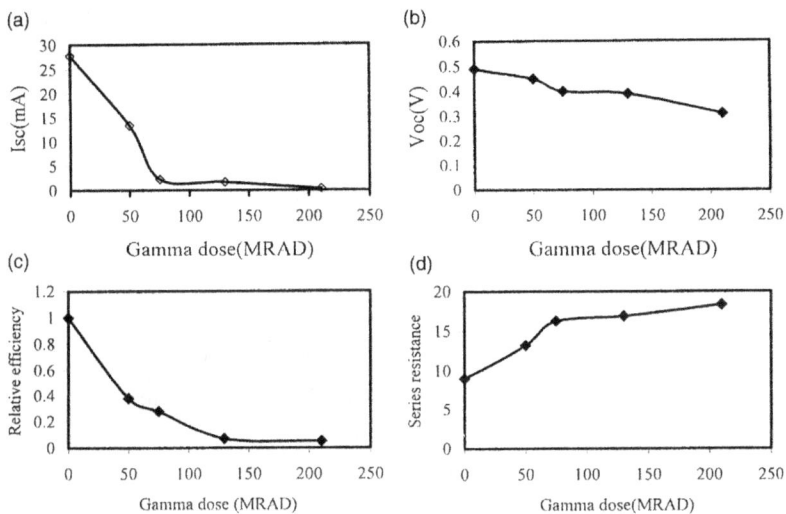

Figure 22   I-V characteristic after different gamma radiations (in MRAD): a) J_{sc}, b) Voc, c) PCE, d) Resistivity [160] (with permission to reproduce/reprint from the publisher).

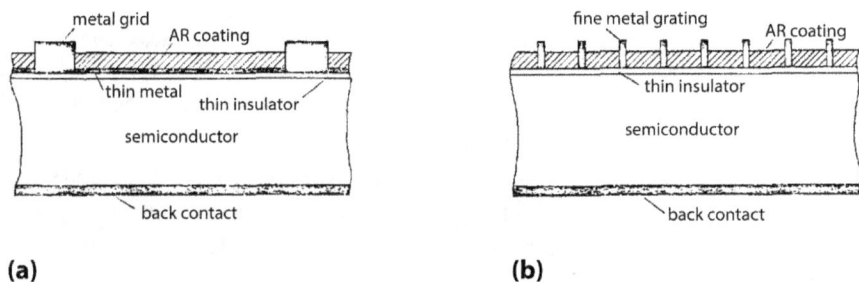

Figure 23   a) Transparent metal method, b) grating method for MIS solar cells [3] (with permission to reproduce/reprint from the publisher).

25

The parameters in Table 6 [164] which have an important role in MIS solar cell optimization are: the substrate resistivity, insulator charge, insulator thickness (< 20Å), diffusion length, l, surface-state density, D_{DL}, J_{ph}, P_b, P_u, crystal orientation, minority-carrier lifetime and so on [130, 164].

Table 6   *Parameters used in MIS solar cell calculations [164] (with permission to reproduce/reprint from the publisher).*

Parameter	Description	Value
P_b	Density of holes	$3 \times 10^{15} cm^{-3}$
$\beta = q/kT$	Inverse thermal voltage	$40 V^{-1}$
$\bar{c}$	Thermal velocity of electrons	$10^7 cm \cdot s^{-1}$
σ_P	Capture cross section of DL	$10^{-15} cm^2$
P_u	Linear density of filled traps	$3 \times 10^7 cm^{-1}$
l	Length of dislocations	$0.05 - 1 \mu m$
D_{DL}	Density of dislocations	$0 - 10^9 cm^{-2}$
E_1	Depth of dislocation level	$0.05 eV$
α'	Geometric factor	0.3
J_{ph}	Photocurrent density	$11.4 mA \cdot cm^{-2}$

Figure 24 [3] shows that PCE decreases appreciably for insulator thicknesses > 20Å, because a large degree of photocurrent suppression occurred in the I-V characteristics. Figure 25 [3] shows the PCE as a function of the insulator thickness. Also, it is found that a MIS solar cells doping increase causes a high PCE [3].

It is found that the metal-insulator-n-semiconductor solar cell has different response to thickness of the insulator in dark and illuminated conditions. In light condition, the ideality factor, n, decreases for thickness > 20Å, whereas in dark condition, n increases for the thickness > 20Å (Table 7) [165].

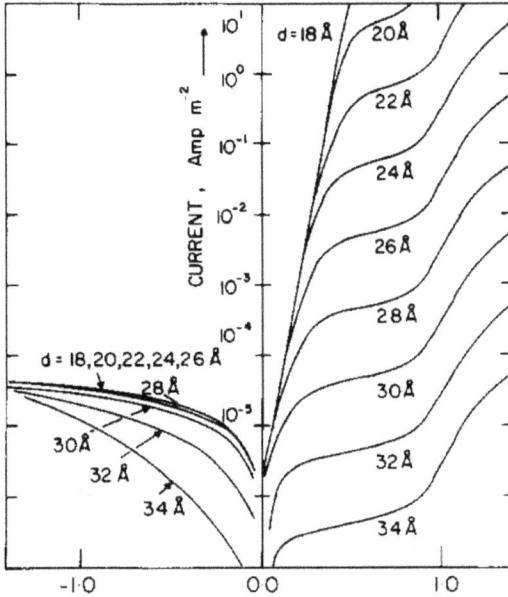

Figure 24   *Significance of the thickness of the i-layer in MIS solar cells [3] (with permission to reproduce/reprint from the publisher).*

Table 7   *Variation of thickness, d, in MIS solar cells; ϵ_i is the dielectric constant and n is the ideality factor [165] (with permission to reproduce/reprint from the publisher).*

Parameters for dark conditions		Parameters for light conditions	
d (Å)	n	ϵ_i (eV)	n
0	1	0.3	1
5	1.1	0.259	1.08
10	1.29	0.2178	1.19
15	1.54	0.177	1.25
20	1.87	0.136	1.266
25	2.41	0.087	1.2
30	3.40	0.054	1.2

For example, for an Au-SiO$_2$-n-Si solar cell under light, n increases with increase of the thickness, while, n decreases for d >18Å [166].

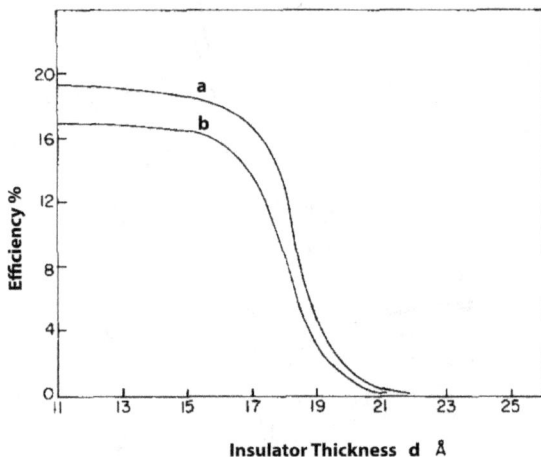

Figure 25   *The effect of insulator thickness on the PCE of the solar cells for two typical values of substrate resistivity (0.07 and 2.0 Ω) [3] (with permission to reproduce/reprint from the publisher).*

One of the important parameters of the MIS solar cells is the barrier height of the MIS junction that can be expressed as follows [3] (Eq. 10),

$$\phi_p = E_{gs} + \phi_{si} - \phi_{mi} - F_s d + \frac{q(Q_{ss}+Q_i)d}{\epsilon_i} \tag{10}$$

Where, Q_{ss} is the charge of the surface states, and Q_i is the charge of the insulator at the interface. F_s is the field strength at the interface, ϵ_i the dielectric constant which usually, ϵ_i of the insulating layer, and d is the interfacial layer thickness [3, 162].

In MIS solar cells, if a doping density and absorption coefficient are below $10^{15} cm^{-3}$, the efficiency loss because of the image force effect is important. Also, for a surface recombination velocity lower than $10^6 cm/sec$, the total loss because of the image force effect and the diffusion model must be compared. The PCE is increased by the thickness optimization, decrease of the surface recombination velocity, design of the high-low doping profile or semiconductor doping to maximize the total solar spectrum [162]. But the PCE performance of MIS solar cells can be deteriorated after high illumination and temperature heat treatment, e.g. isothermal and isochronal preprocessing or hydrogen plasma treatment due to decrease of carrier concentration and less J_{sc} and Voc. These phenomena increase the recombination in surface heavily occupied by electrons which increase the density and decrease the barrier height and the PCE. So, it is important to locate them in rapid heat treatment with low temperature to improve the cell parameters [167-172] (Figure 26 [170]).

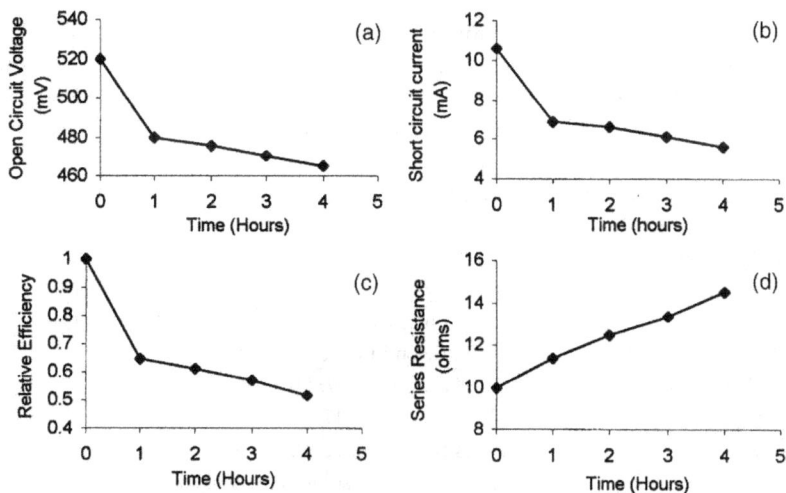

Figure 26   Impact of effect of high treatment on the output parameters of MIS solar cells: (a) Voc, (b) J_{sc}, (c) PCE and (d) resistivity [170] (with permission to reproduce/reprint from the publisher).

MIS solar cells are also cost competitive due to having major benefits such as low temperature process, improved response, high Voc, reduction of shunting, simple fabrication and excellent structure [173, 174]. They also have high PCE and because of this characteristic, they can be used instead of p-n junction solar cells. Using the chromium as a SB increases the PCE of large MIS solar cells to 12% on single-crystal Si and over 9.5% on Wacker polycrystalline Si [149, 175, 176]. They are reused by a new technique as a defective MIS etched in H_3PO_4 acidic solution. In this technique all insulating layers – i.e. silicon oxide and silicon nitride – are etched after about 30 min [177].

3.11.2 Semiconductor-insulator semiconductor (SIS)

SIS solar cells have exact characteristic which make them better than SB, MIS, and heterojunction structures, e.g., Si inverted at the hybrid interface with the polymer creating an SIS without any significant insulating film [178], for solar energy conversion. These solar cells are more stable and efficient than MIS or SB solar cells. The potential superiority of SIS is that it does not have any thin metal, junction and interfacial layer which absorbs the light caused environmental degradation problem and minimizes the effect of the interface states and band gaps and conductivities in semiconductor as the top layer. In fact, The SIS structure requires a thin SiO_2 interfacial layer with a thickness lower than 18Å (> 10Å) existed between the top layer and the Si which is done by controlled furnace tubes or boiling water. So, the top layer acts as a low-resistance window, an AR coating and the collector layer of the junction. Also, SIS with wide-gap semiconductors removes the dead layer of the surface in homojunction devices which improves the UV responses [33, 179-182]. The top layers of n-type and p-type SIS solar cells are provided with conducting layer with low electrical resistivity, which must be transparent for light to reach the semiconductor, instead of the metal and currently 12% of them use oxide semiconductors, such as $(In_2O_3)_{0.91}(SnO_2)_{009}$, SnO_2, SnO_2:As, ZnO, ITO and tin antimony oxide on Si which ITO and SnO_2:As are better, because SIS matches the main properties of the ITO on Si, e.g. ITO-SiO_2-Si, and it can be used for n-type and p-type SIS solar cells [3, 7, 40, 59, 147, 179, 183-188]. Also, as previously introduced, ITO has many applications here. Semiconducting metallic oxides such as ITO have the unique characteristic as top layers in SIS solar cells. They have great transparency, great conductivity, and superior environmental stability which offer electrical and optical parameters in contrast to simple oxides, e.g., tin oxide or indium oxide. Because the optical absorption edge changes with mixing of alloying or bonding between that of indium oxide and tin oxide [189] or such an absorption caused by the geometric composition of the materials [190]. Some effective properties of ITO are as follows [179]:

1) ITO acts as a metallic layer and degenerate n-type semiconductor to decrease series resistance in the top layer films in the SIS structure.

2) n_{AR} in ITO [191] is ~2.0 to make it a partial AR coating on Si.

3) The ITO layer films are resistant to chemical attack, stable under most environmental conditions, and adhere to many materials well.

4) The wide band gap of ITO is 3.6eV which is wide with minimal visible absorption in the oxide layer. So, it passes most of the light through it.

5) ITO can be deposited at low temperature, so that abrupt junctions can be shaped and interdiffusion is reduced. Also, the effective "zero junction depth" decreases any surface

dead layer and enhances short-wavelength response. But, the surface states at the interface with insulator layer must be minimized.

Both ITO and tin antimony oxide have significant photovoltaic PCE on Si substrates. Table 8 shows some typical properties of ITO layer films [179].

Table 8   *Typical characteristic of ITO layer films [179] (with permission to reproduce/reprint from the publisher).*

Crystal structure	Cubic
Lattice constant	10.118 Å
Linear coefficient of thermal expansion	$10.2 \times 10^{-6}/°C$
Band gap	2.62-3.75 eV
Resistivity	$1.77 \times 10^{-4}\ to\ 1.2 \times 10^{-2}\Omega\ cm$
Carrier concentration	$1 \times 10^{19}\ to\ 1.3 \times 10^{21} cm$
Effective mass	$0.3m_0\text{-}0.55m_0$
Hall mobility	$26\text{-}40\ cm^2/V\ sec$
Refractive index	1.68-2.48
Electron affinity	4.1-4.3 eV
Position of Fermi level below or above the CB edge	0-0.4 eV

New SIS solar cells utilize ITO at the top layer and polycrystalline Si as the base layer. These types of SIS solar cells are used for large-scale photovoltaic applications [22, 57, 192, 193]. The I-V characteristic of three ITO/polycrystalline-Si SIS solar cells are shown in Figure 27 [57].

As shown in Figure 27 [57], chemical treatment of the surface has improved the diode characteristic. Also, the FF is low, because no grid was employed to them. The recombination and electrostatic role of the surface states have decreased the hydrogen passivation. So, Voc has been improved and it shows that the hydrogen improves the Voc due to reduction of the electrostatic effect [57], but small diodes enhance the leakage current problems up to the large-area diodes [194].

In SIS solar cells, when the metal is replaced with a degenerate wide-bandgap oxide semiconductor, it performs as same as MIS. It has also the same performance as p-n junction solar cells with controlling the interfacial layer thickness and the true selection of oxide semiconductor [40].

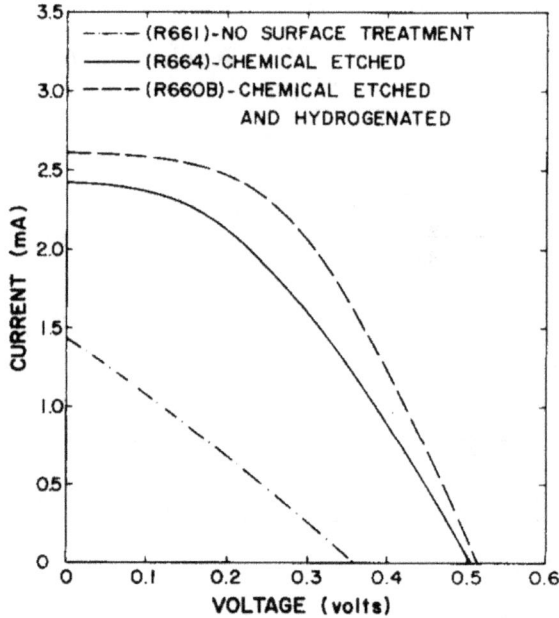

Figure 27   I-V characteristics of ITO/polycrystalline-Si SIS solar cells for different surface treatments [57] (with permission to reproduce/reprint from the publisher).

The major parameters of the SIS solar cells are the oxide semiconductor work function and the insulating layer thickness. Many defects can be present because of the lattice constants, thermal expansion coefficient, and mismatch of the crystal structures. Table 9 [40] shows the most mixing of the top layer semiconductor and base layer semiconductor. Sections (a) and (b) in Figure 28 [40] show the diagram of the simple equilibrium energy-band for n-oxide-semiconductor/p-base-semiconductor solar cells, with and without the existence of an interfacial layer [40].

Table 9   *Some characteristic of the top layers and base layers in SIS solar cells [40] (with permission to reproduce/reprint from the publisher).*

Semiconductor	Type of crystal structure	Lattice constant (Å)			Linear coefficient of thermal expansion $(10^{-6}/°C)$	Band gap (eV) at $300°K$	Electron affinity (eV)
		a	b	c			
Si	Diamond	5.431	-	-	2.3	1.12	4.05
Ge	Diamond	5.657	-	-	5.8	0.66	4.0
GaAs	Zinc blende	5.653	-	-	5.9	1.43	4.07
InP	Zinc blende	5.869	-	-	4.5	1.34	4.4
CdTe	Zinc blende	6.477	-	-	5.9	1.44	4.3
$CuInSe_2$	Chalcopyrite	5.782	-	11.620	-	1.02	4.15
In_2O_3	Cubic	10.118	-	-	10.2	~3.0	4.3-4.4
SnO_2	Tetragonal	4.737	-	3.185	4.0	~3.5	4.8-4.9
ITO (90% In_2O_3+10% SnO_2)	Cubic	10.118	-	-	~10.2	~3.65	4.1-4.5
ZnO	Hexagonal	3.249	-	5.205	7.2	~3.2	4.2
CdO	Cubic	4.694	-	-	-	~2.7	4.47
Tl_2O_3	Cubic	10.543	-	-	-	~2.3	4.21
Bi_2O_3	Cubic	10.245	-	-	-	~2.9	4.52
Cd_2SnO_4	Orthorhombic	5.568	9.887	3.902	-	~2.9	4.55
In_2TeO_6	Hexagonal	8.883	-	4.822	-	-	-
MoO3	Rhombic	3.90	13.8	3.7	-	~3.75	4.58

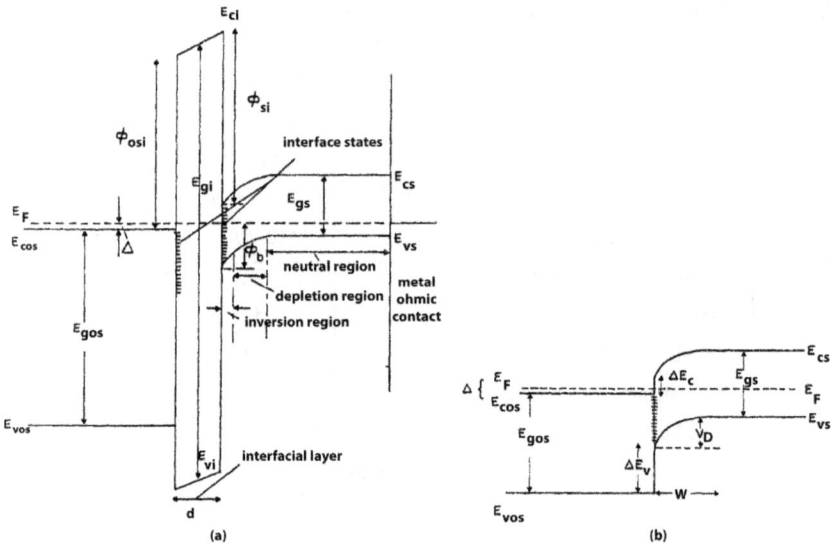

Figure 28   *(a) The diagram of a simple equilibrium energy-band of SIS solar cells. E_{gi}, and E_{gs} are the insulator, and base semiconductor, respectively. ϕ_{osi} is the oxide semiconductor-to-insulator barrier height related to the oxide semiconductor work function. V_i is the potential drop across the thickness interfacial layer, d. ϕ_b is the device barrier height, (b) The diagram of a simple equilibrium energy-band for n-oxide semiconductor/p-based semiconductor solar cells, without the existence of an interfacial layer [40] (with permission to reproduce/reprint from the publisher).*

In oxide semiconductor/base semiconductor structures, the interfacial layers are arranged with two types: 1) deposited dielectrics, or 2) native interfacial layers - i.e., commonly an oxide - grown to the base semiconductors. According to the energy diagram at Figure 28(a), the most crucial parameter for choice of interfacial layers is E_{gi} which is as follows (Eq. 11) [40],

$$E_{gi} \geq (\chi_s - \chi_i + E_{gs}) \tag{11}$$

Where χ_s is the barrier height for the surface states and χ_i is that for the interfacial layers. However, considering the studies, it has been shown that SnO_2 is the only candidate for n-type base semiconductors and the ZnO and ITO are only proper candidates for p-type base semiconductors [40]. SnO_2:As – i.e. arsenic-doped tin oxide films made textured and produced by Chemical Vapor Deposition (CVD) method [52, 195] – improves the PCE over 14.5% and FF, Voc and J_{sc} on single Si crystal [188].

Another research has shown that using ITO-SiO$_2$ in the top layer and Si in the base layer at SIS solar cells, known as ITO-SiO$_2$-Si SIS solar cells, results in high current and low Voc with PCE of 16% for active area, and 14% for total area [196].

3.11.3 Electrolyte-insulator-semiconductor (EIS)

In EIS solar cells, the work function of the metal is replaced with the work function of the electrolyte (Figure 29) [3].

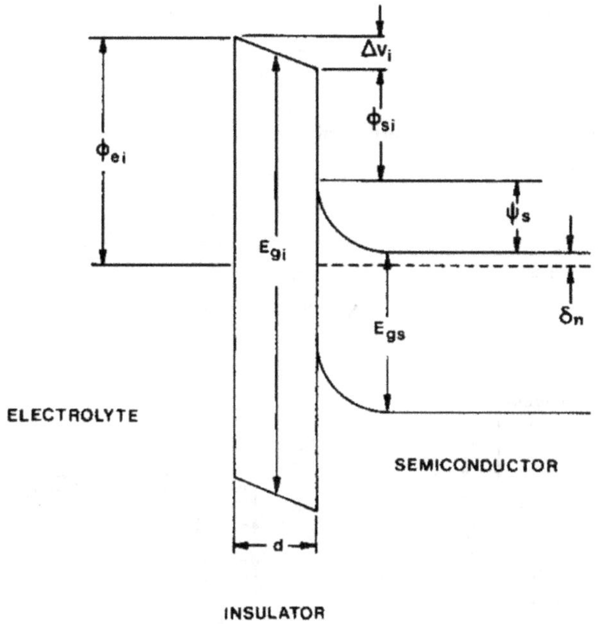

Figure 29   The diagram of EIS solar cells [3] (with permission to reproduce/reprint from the publisher).

Figure 29 gives the Eq. 12 for EIS solar cells [3],

$$\phi_{ei} = \Delta V_i + \phi_{si} + \psi_s + \delta_n + \frac{q(Q_{ss}+Q_i)d}{\epsilon_i} \tag{12}$$

Where, ψ_s is the surface potential of the semiconductor, Q_{ss} is the charge of the surface states, and Q_i is the charge of the insulator at the interface. The ϵ_i is the dielectric constant which usually, ϵ_i of the insulating layer, d is the interfacial layer thickness, and δ_n is the distance between the Fermi level in the bulk and the CB of the semiconductor bulk [3, 162].

EIS solar cells are new. At first, they are produced on (100) n-type tellurium-doped GaAs wafer which carrier concentration is $5 \times 10^{22} m^{-3}$. Then, these wafers are etched for 15 seconds and then etched again for 1 minute. After etching, they are washed with deionized water and afterwards dried with nitrogen. After that, oxides are grown on them by low temperature and wet oxygen. Then, they are embedded in a quartz tube to be saturated with water vapor for 50 hours. They are uncovered in electrolyte, composed from a 1/1 mix of AlCl₃/butyl pyridinium chloride (BPC). They can then be used for the photovoltaic output of MIS and SIS solar cells, too, because existence of an ultrathin tunable interfacial layer on the electrode surface can prevent solar cells from photo electrochemical corrosion [3, 197].

3.12   Conductor-insulator-organic semiconductor (CIOS)

A novel semiconductor, called CIOS, have the characteristic of both thin insulating layers and transparent conductive front electrodes in an organic solar cell. In CIOS, three kinds of CIS structures including ITO/X-H₂Pc/Au, ZnO/X-H₂Pc/Au and CdS/X-H₂Pc/Au are used that the ZnO/X-H₂Pc/Au presents considerable long-time stability. Figure 30 [147] shows two types of CIOS such a sandwich cell which has an organic semiconductor.

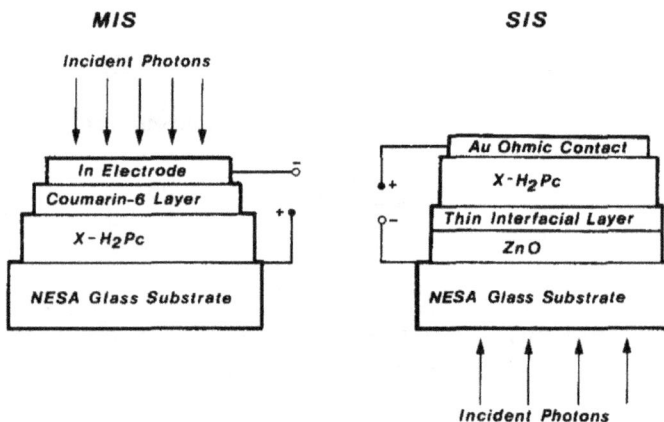

Figure 30   Two types of CIOS [147] (with permission to reproduce/reprint from the publisher).

Table 10 [147] shows the characteristic of X-H₂Pc based CIOS solar cells. Optimization of the contacts and parameters of the semiconductors can improve the cells. This optimization leads to increase of Voc and J_{sc} of devices [147].

Table 10   Characteristic of X-H₂Pc based CIOS solar cells [147] (with permission to reproduce/reprint from the publisher).

Structure type	Conducting layer	Voc (V)	J_{sc} $(\mu A \cdot cm^{-2})$	FF
SIS	SnO₂	0.035	5.7	0.3
SIS	In₂O₃	0.21	145	0.33
SIS	ZnO	0.45	176	0.45
SIS	CdS	0.62	130	0.30
MIS	In/Coumarin	0.39	186	0.35

When the conducting layer is ZnO, the photovoltaic effects stems from light absorption in X-H₂Pc based CIOS solar cells. Researchers have also shown that CIOS solar cell are more stable and efficient than MIS solar cells, but the PCE of CIOS solar cells are still very low and researchers try to improve it for practical device [147].

3.13   Metal-organic-insulator semiconductor (MOIS)

MOIS such as alkyl chain molecules on n-Si are created by hitching alkyl chain molecules, with sulfur termination etc. which limits the inversion on Si or methyl-terminated monolayers, via Si-O-C bonds to oxide-free n-Si surfaces and Au nanoparticles via electroless plating to top contacts, e.g. Au – i.e. n-Si-O-C₁₁-S–Au, Hg, n-Si-O-C₁₁-S-Hg – without any chemical bond to the molecules (Figure 31) [198]. This semiconductor has 25 mW/cm^2 white light illumination, Voc = 0.48-0.54V and FF = 0.58-0.8%. The PCE of these types of semiconductors is very low and researchers follow a solution by using the proper transparent conductors on top of the methyl-terminated monolayers to enhance their PCE [198].

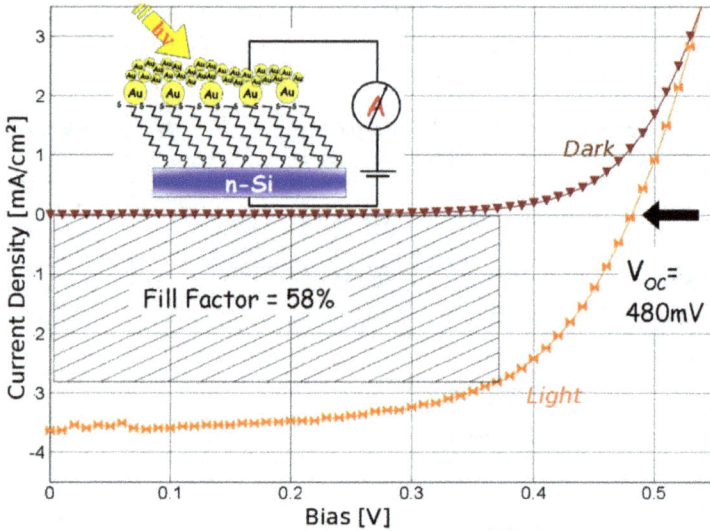

Figure 31   I-V characteristic of MOIS [198] (with permission to reproduce/reprint from the publisher).

3.14   Metal-insulator-nonuniformly-doped semiconductor (MINDS)

Metal-insulator-nonuniformly-doped semiconductor (MINDS) solar cells, e.g., four layered metal-insulator-pp+ semiconductor MIS solar cell, are semiconductor layers which are doped nonuniformly in which drift electric field is created. In these solar cells, a p-type layer such a p+-substrate is grown. By increase of temperature, the impurity atoms, N, distribute from the epitaxial layer and this impurity is controlled by the diffusion time and temperature (Eq. 13) [199],

$$N(x) = N_B + \frac{N_A}{2}\operatorname{erf}(bx) \tag{13}$$

Where N_B is the initial epitaxial layer doping density, N_A is p-substrate doping density, x is semiconductor distance and b is the doping gradient (Eq. 14) [199],

$$b = \frac{1}{2\sqrt{D_d \tau_d}} \tag{14}$$

Where τ_d is the diffusion time as well as D_d is the diffusion coefficient of the acceptor impurities.

Figure 32 [199] depicts the doping density, inside the p-type epitaxial layer, where the doping is assumed to be uniform inside the substrate p+ region. Table 11 [199] depicts the parameters of this solar cell.

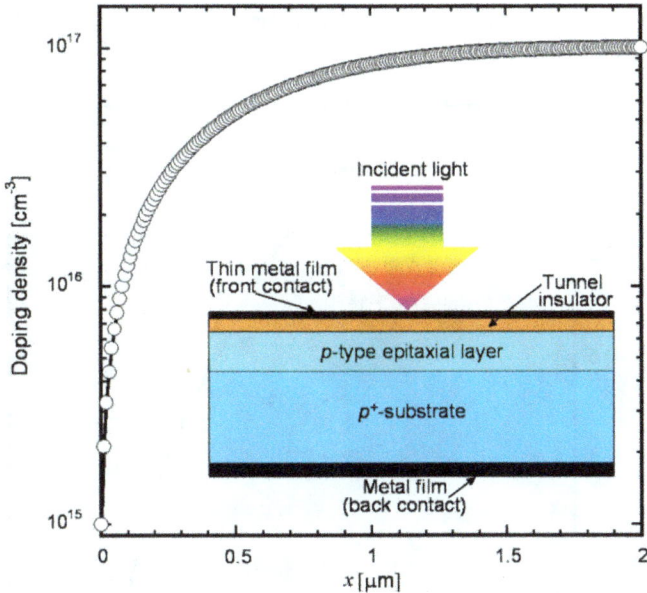

Figure 32   Doping density inside the p-type epitaxial layer [199] (with permission to reproduce/reprint from the publisher).

Table 11   Parameters of MINDS solar cells [199] (with permission to reproduce/reprint from the publisher).

Parameter	Description	Value
N_A	Doping density in the p+-substrate	$1 \times 10^{17} cm^{-3}$
N_B	Initial doping density in the epitaxial layer	$1 \times 10^{15} cm^{-3}$
b	Doping gradient in the epitaxial layer	$1 \times 10^4 cm^{-1}$
d	Insulator thickness	$12Å$
L	Semiconductor layer thickness,	$125\mu m$
Q_s	Fixed insulator charge	$1 \times 10^{12} cm^{-2}$

Figure 33 [199] shows the calculated spectral response of this solar cell at various temperatures which is moved to higher wavelength with high temperature referred absorption of the semiconductor bandgap due to increase of the temperature and minority carrier diffusion length in the substrate area. The high temperature leads to reduction of the Voc and PCE. Thus, lower temperature is better for this solar cell performance, and their maximum PCE is 10% [199, 200].

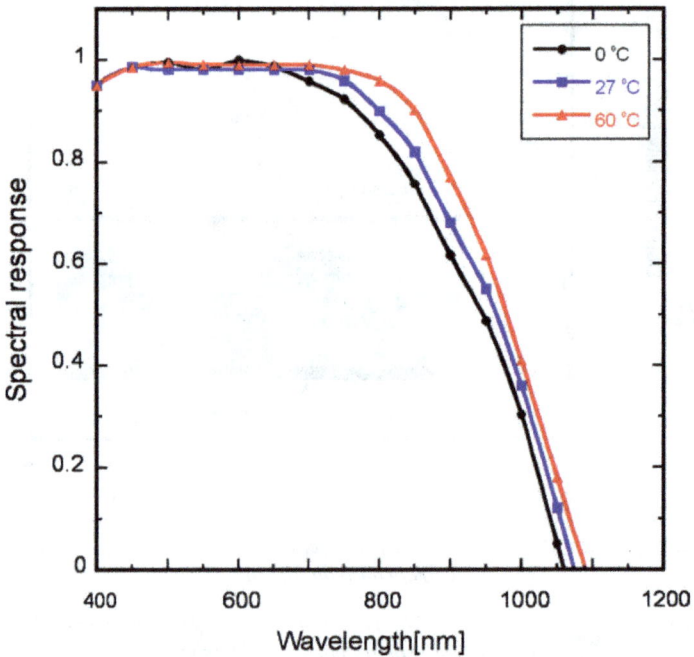

Figure 33   The calculated spectral response of this solar cell at various temperature [199] (with permission to reproduce/reprint from the publisher).

3.15   Metal-nanowire-network-metal-insulator semiconductor (MNN-MIS)

The MIS structure is created with a thin oxide and detrimental metal contact with the semiconductor base. This semiconductor solar cell has properties of p-n junction solar cells, and has simplicity of fabrication, and relatively large Voc. It has one major disadvantage as all MIS structures that a specific metal layer is required to shape a junction at the Si surface for increasing the optical transmittance and J_{sc}. This drawback can be resolved by nanoscale structures such that the nanowire metal networks can provide that for the MIS. In this technology, it is used conductive and adjacent metal wires to generate an inversion layer for a high optical transparency and homogeneous CC extraction which provides high Voc and J_{sc} as well as PCE of ~11% [201]. Figure 34 [201] shows the Au−Pd metal nanowire MIS, and Figure 35 [201] depicts the I-V characteristic of that.

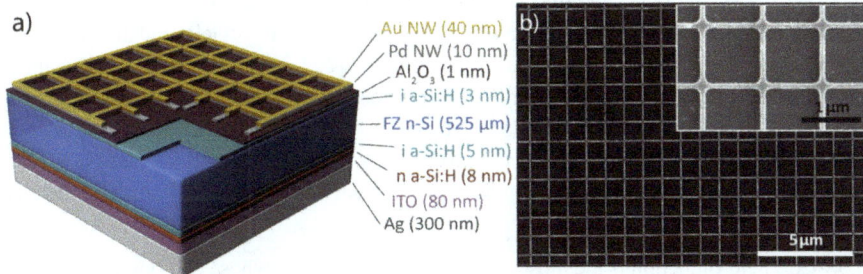

Figure 34   *a) Au−Pd metal nanowire MIS with various layers, b) Scanning Electron Microscope (SEM) images of the nanowire network with 50 nm height, 100 nm width, and 1 μm pitch on top of the silicon half-cell [201] (with permission to reproduce/reprint from the publisher).*

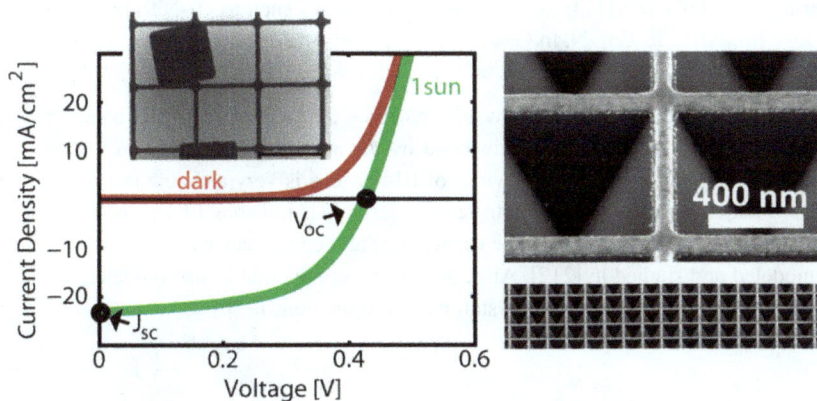

Figure 35   *The I-V characteristic of Au-Pd metal nanowire MIS [201] (with permission to reproduce/reprint from the publisher).*

3.16   Polycrystalline compound semiconductor (PCS)

Polycrystalline compound semiconductor (PCS) solar cells, specially III–Vs, II–VIs, I–III–VI$_2$s [66], II-VI, I-III-VI$_2$, and IIO$_x$VI$_{1-x}$mixtures, such as CuInSe$_2$, CuGeSe$_2$, and also CdTe and Si compounds, are cheap photovoltaics. The PCE of them is more than 12% and reaches 15% with high minority carrier lifetime and excellent resistivity. Other PCSs are InP, InGaP, InGaAsP, GaAs, AlGaAs, InGaAs, Al$_{0.35}$Ga$_{0.65}$As, Al$_{0.25}$Ga$_{0.75}$As, In$_{0.15}$Ga$_{0.85}$As, Si and Ge compounds, CuInSe$_2$, etc. that InP-based solar cells, e.g., InP, InGaP, and InGaAsP, and CuInSe$_2$ solar cells have high radiation resistance [202-204]. Si compounds, e.g., Si Tetrachloride (SiCl$_4$), Silane (SiH$_4$), or Si-chloro-hydrogen compounds are called the volatile compounds cleaned by chemical methods. Other Si compounds are quartz (SiO$_2$), Si$_3$N$_4$, SiC, and combination of them grown by Czochralski-method. PCS solar cells such as GaAs, InP-based, CdSe, ZnP, Cu$_2$S, and Cu$_2$O solar cells are very expensive and not cost-efficient [202].

CU$_2$S-CdS heterojunction is the most prominent PCS solar cell whose compound structure is p-type or n-type [48, 205]. The pn-heterojunctions for PCSs have shown in Figure 36 [205].

In PCSs, the CB alignment is particularly important to increase Voc and J_{sc}. Hence, they need to be optimized in accordance with all crystallographic orientations [205].

Other PCSs, as known III-V compound semiconductors such as chalcopyrite-series CIGSSe, kesterite-series CZTSSe [174, 206-209] and, their alloys such as III-Nitrides and III-N-V, e.g., InN, In$_{1-y}$Al$_y$N, In$_x$Ga$_{1-x}$N, In$_y$Ga$_{1-y}$N$_x$As$_{1-x}$, absorb a wide range of solar radiation which is very valuable for solar cells and gives them high-grade PCE [53, 204, 210, 211].

For example, In$_x$Ga$_{1-x}$N, e.g. In$_{0.22}$Ga$_{0.78}$N, receives the maximum of the solar irradiance $5 \times 10^4 cm^{-1}$ which is effectively absorbed by the solar cells and generates photovoltaic energy [18, 19, 211]. Also, the behaviour of III-Nitrides is very amazing due to their great PCE and better resistance in different temperatures [211]. Table 12 shows the simulated parameters of In$_{0.22}$Ga$_{0.78}$N [211]. The capacitance and conductance of such heterojunctions was modeled and studied in [212]. Also, the current conduction is also modeled and an I-V relation was derived for such poly crystalline heterojunctions [73].

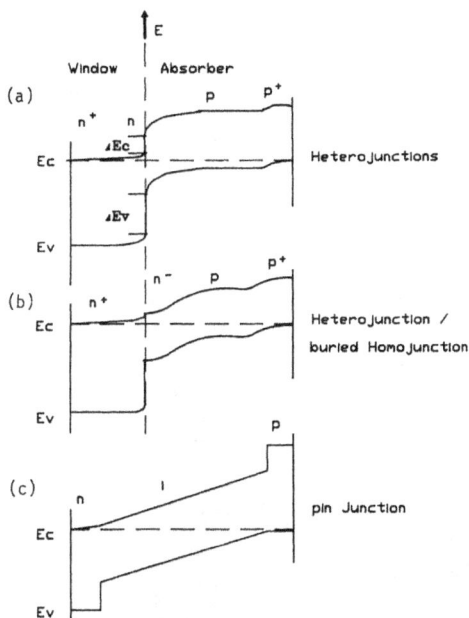

Figure 36   The pn-junctions for PCSs [205] (with permission to reproduce/reprint from the publisher).

Table 12   Simulated parameters of the In$_{0.22}$Ga$_{0.78}$N semiconductor [211] (with permission to reproduce/reprint from the publisher).

Simulated parameters	Description	Results from simulated parameters	Obtained values
Doping of front (cm^{-3})	3.9×10^{18}	Current $J_{sc}(mAcm^{-2})$	49.1
Doping of base (cm^{-3})	5×10^{17}	Voltage Voc (V)	1.2
Front thickness (μm)	0.3	FF	0.8
Base thickness (μm)	20	Maximum power ($mW cm^{-2}$)	47.13
Solar lighting	AM 1.5	η (%)	34.6

3.17   Microcrystalline semiconductor (MicoSem)

Microcrystalline semiconductor (MicoSem) solar cells, especially ultrathin and III-V mixtures, such as GaAs/AlGaAs, have p-i-n diode structures with intrinsic layer. The undoped layer thickness of these solar cells is usually $10\mu m$ and their PCE is about 12-15%, so they have commercial applications [52, 202]. Ultrathin, is a MicoSem which have many applications in solar cells [213]. III-V compounds have many applications in Medium Concentrator Photovoltaic (MCPV) markets [214]. Figure 37 [213] shows a scheme of the MicoSem as a multilayer epitaxial compound of GaAs/AlGaAs.

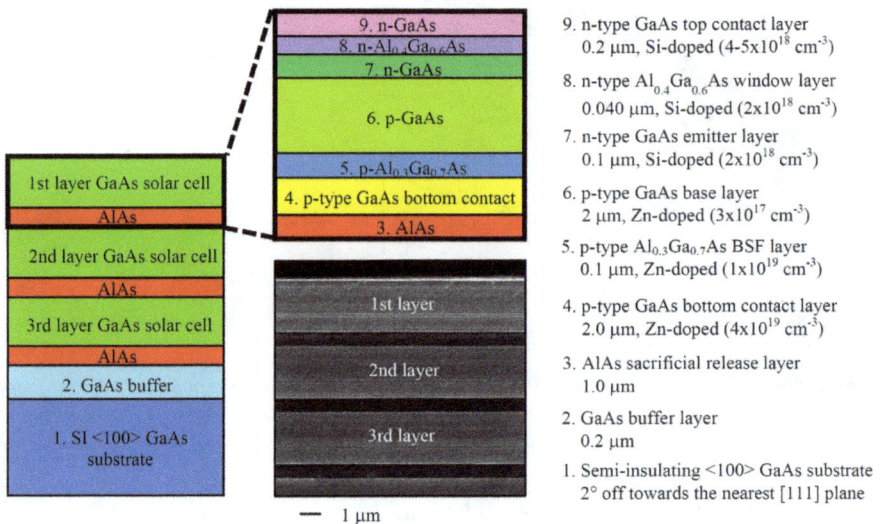

Figure 37   Scheme of GaAs/AlGaAs MicoSem [213] (with permission to reproduce/reprint from the publisher).

The MicoSems are fabricated by growth of epitaxial layers on a reusable substrate. This technology can also be used for lasers, sensors, light emitting diode, and other materials system. One of the benefits of these semiconductors is cost which makes them available for conventional technologies [213]. A new study has shown that GaAs/AlGaAs alloy is not proper to utilize at temperatures less than 110°K and higher than 540°K due to J_{sc} reaching zero, and weak PCE [215].

3.18 Nanocrystalline semiconductor (NS)

Nanocrystalline semiconductor (NS) solar cells, such as CIGS [216], Al_2O_3 coated TiO_2 films and an I_2/NaI doped solid-state polymer electrolyte, have high PCE due to their sizes, maximum $20nm$, which reduce the scattering rate and increase the state density. So, by changing the size of the crystallites, it can be feasible to tune the band gap for absorbing the energy. Under $10mW \cdot cm^{-2}$ AM1.5 illumination, they show high PCE of $\sim5.3\%$. CGIS can be a good alternative for poly Si semiconductors [216-218]. NSs exist in such an array or single structure [91]. A reason for invention of NS solar cells is that the thin film solar cells are nonuniform because of their fabrication in large area, so NS solar cells do not have this problem as well as they are very cheap. They are called low-cost solar-energy-harvesting devices which have simple fabrication [109, 217-219]. The NS solar cells are made by electrochemical synthesis which uses thermal decomposition process, nonaqueous sol-gel process, and other methods such as hydro/solvothermal process, polyol synthesis and the aqueous solution, solution phase method, etc. [91] or dye synthesis [220] of, e.g. CdS, ZnS, TiO_2 [91, 221], Sb_2S_3, CuSCN [221], Boron-doped copper indium telluride (B-doped $CuInTe_2$) [222], solid-state polymer electrolyte inside pores [217, 218, 220], etc. which sensitizes the N3, N719 [220], $AgBiS_2$ [223], SnO_2 electrode [224] and photocurrent action spectrum [225]/Luminescent Down-Shifting (LDS) [226] effect onto surfaces such as ITO [225], or CdS, CdSe, CdS_xSe_{1-x}, $Cd_xZn_{1-x}S$, InP, AlInP [227], CdTe, CdHgTe, InAs, and PbS [219] and GaAs quantum wires [226], as nanowires/NPs/Quantum Dots (QDs), on e.g. aluminum (Table 13 [109]) [109, 228-230]. They are flexible in accordance with the substrate and semiconductor materials [49, 218, 231]. The NS solar cells made by dye synthesis method, are called Dye-sensitized solar cells (DSSCs) which is shown in Figure 38 [108, 220, 232, 233]. A novel method uses organic-inorganic lead halide based perovskite as a sensitizer which is very similar to the dye synthesis method [68, 230-236]. The most conventional nanofabrication methods are chemical vapor deposition, epitaxial material growth, and electron-beam deposition. But these methods are very expensive. Therefore, researchers try to develop other cheaper techniques, for example using organic polymers, and microcrystalline semiconductors, self-organized, and electro-deposition [237-242], but these methods cannot have essential control over size distribution. A novel method which has $\pm10\%$ control is performed by electrochemical anodization of aluminum which is cheap. This technique can be utilized in multijunction structure too as demonstrated in Figure 39 [49] for increasing the PCE [49, 243].

Table 13   *Quantum wire solar cell summary [109] (with permission to reproduce/reprint from the publisher).*

Group	Material	Structure	Voc (V)	J_{sc} (mA/cm^2)	FF (%)	η (%)
IV	Si	Single	0.29	3.5	0.51	0.5
IV	Si	Single	0.26	23.9	0.55	2.3-3.4
IV	Si	Single	0.19	5	0.4	0.46
IV	Si	Array	0.29	4.28	0.33	0.46
IV	Si	Array	0.549	26.06	0.651	9.31
IV	Si	Array	0.75	17.5	0.45	5.9
IV	Si	Array	0.525	16.45	0.559	4.83
IV	Si	Array	0.13	1.6	0.28	0.1
IV	Si	Array	0.339	3.5	0.36	2.7
IV	Si	Array/Pt QD	0.55	20.2	0.61	8.14
III-V	GaN	Single	1	0.39	0.56	0.19
III-V	GaN	Array	0.95	7.6	0.38	2.73
III-V	GaAs	Array	0.2	0.201	0.267	0.83
III-V	GaAs	Single	1	<0.1	0.65	4.5
III-V	InAs	Array	0.75	0.2	0.53	2.5
III-V	InP	Array	0.43	13.72	0.57	3.37
SCSSC	ZnO	Array/CdSe QD	0.6	2.1	0.3	0.4
SCSSC	ZnO	Array/CdSe/CuInS$_2$ QD	0.45	3.21	0.49	0.71
SCSSC	ZnO	Array/Cu$_2$O NP	0.15	1.43	0.25	0.053
SCSSC	ZnO	Array/Cu$_2$O	0.29	8.2	0.36	0.88
SCSSC	TiO$_2$	Array/Sb$_2$S$_3$ QD	0.556	12.3	0.699	5.06
SCSSC	CdS	Array/CdTe	0.62	21	0.43	6
Hybrid	CdS	Array/Polymer	0.6	5.26	0.54	1.73
Hybrid	CdSe	Array/Polymer	0.85	2.96	0.47	1.17
Hybrid	CdSe	Array/Polymer	0.7	6	0.44	1.7
Hybrid	CdTe	Array/Polymer	0.714	3.12	0.477	1.06
Hybrid	ZnO	Array/Polymer	0.3	1.73	0.389	0.2
Hybrid	TiO$_2$	Array/Polymer	0.61	11.1	0.66	5
Hybrid	ZnO/TiO$_2$	Array/Polymer	0.5	1.14	0.5	0.29
Hybrid	ZnO/CdSe	Array/Polymer	0.51	4.89	0.35	0.88
Hybrid	GaAs	Array/Polymer	0.53	10.7	0.55	3.1

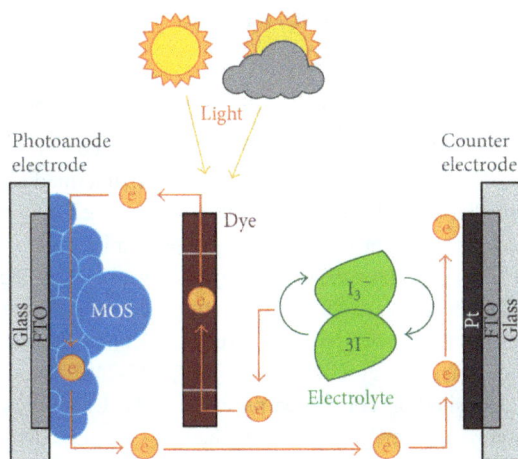

Figure 38   Schema of a DSSC device [232] (with permission to reproduce/reprint from the publisher).

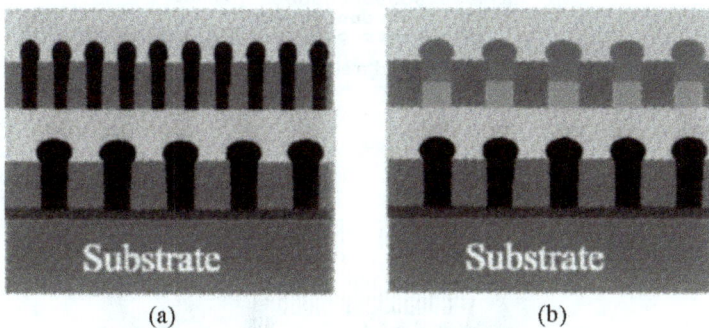

(a)   (b)

Figure 39   NS multijunction solar cells with varying (a) nanostructure size, and (b) nanostructure material compound [49] (with permission to reproduce/reprint from the publisher).

Overall, by controlling the size and mixture of the nanostructures, the absorption characteristic of solar cells can be enhanced for the solar spectrum. So, the technology, as known electrochemical anodization of aluminum, can be used for economic larger area PV cells with different substrate materials which can increase the PCE of these solar cells to 18% and higher [49].

3.19   Quantum dot intermediate band semiconductor (QD-IBS)

The QD intermediate band semiconductor (QD-IBS) solar cell is based on concept of Quantum Ratchet (QR) approach [244] and employed in Intermediate Band Solar Cells (IBSCs). The Intermediate Band (IB) solar cell is dependent on optical and electronic characteristic. These kinds of solar cells work based on semiconductor materials and they have high PCE, 63.2%, for absorbing the low energy photons, i.e., 124-255 meV, and produce high J_{sc} and Voc. They actually show the properties of the carrier density in zero-dimensional nano-structures which creates the band offset between semiconductor materials and dots – i.e. n- and p-emitters called QD, (Figure 40 [245]) [227, 245-248]. They can be with metal nanoparticles [249] deposited on dots called plasmonic QD. Using the metal nanoparticles increases the radiation rate in semiconductor by changing the state density in emitter [227, 230, 250].

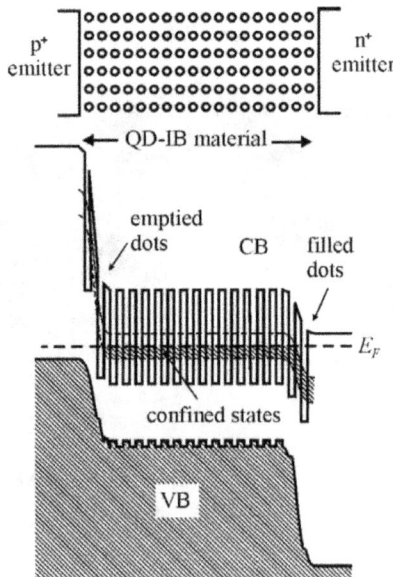

Figure 40   Schema of QD-IBS solar cell, when device is on, pumps electrons from VB to CB, and E_F is Fermi level [245] (with permission to reproduce/reprint from the publisher).

Figure 41 shows the J_{ph} response of QD-IBS and plasmonic QD-IBS [251].

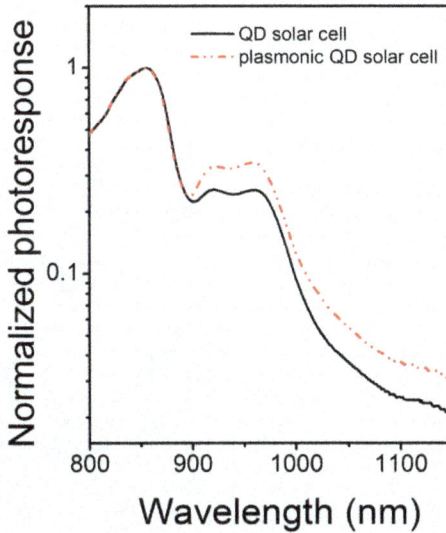

Figure 41   *Photocurrent response of QD-IBS and plasmonic QD-IBS solar cells [251]*
(with permission to reproduce/reprint from the publisher).

QD-IBS solar cells are made based on III–V semiconductor standard like GaAs and InGaN [251, 252] produced by molecular beam epitaxy (MBE)-grown layers [253], e.g., MBE-grown InAs–GaAs structures [245]. Behind the QD technology, there is another technology similar called Quantum Well (QW) technology. Figure 42 [254] compares these two technologies with each other.

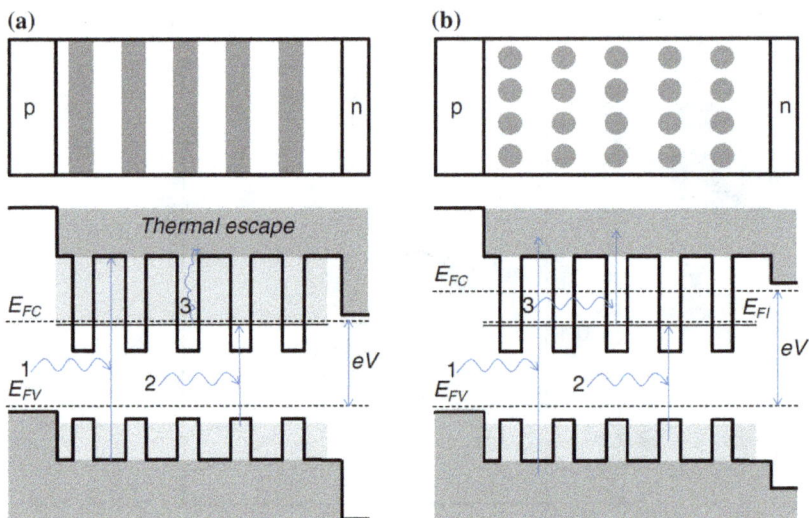

*Figure 42   a) Above, schema of a QW semiconductor and, below, its band; pumping the
electrons from the VB to CB by the photons labelled 1, photons labelled 2 have lower
energy to pump electrons from the VB to the QW band levels. This converts to the current,
because of thermal escape through the barrier material CB. So, the voltage will not
increase the bandgap between the VB and confined levels of QWs. b) Above, schema of a
QD semiconductor and, below, its band; pumping the electrons from the VB to CB by
photon absorption labelled 1, photons labelled 2 have lower energy to pump electrons from
the VB to the IB and photons labelled 3 from IB to the CB. The voltage can reach to the CB-
VB bandgap. Grey tones in the bandgaps depict the non-zero density of states: the lighter
the lower [254] (with permission to reproduce/reprint from the publisher).*

3.20   Multi-type QDs photo-induced doping enhanced semiconductor (MQD-PDES)

In this technology, semiconductors are placed of metal and avoid energy losses by converting
energy into heat [255]. For example, InP and ZnO QDs with band gaps of 2.4 eV and 3.3 eV,
respectively, simultaneously dope Graphene (Gr) by a photo-induced doping method under
UV light. This semiconductor has a great stability and its PCE increases from 8.57% to
11.50%. By choosing correct mixture of materials in an oriented geometry, the PCE of the
solar energy conversion increases [255-257]. Figure 43 [256] shows the VB and CB of this
semiconductor when Gr approaches the semiconductor. The best PCE for MQD-PDES solar
cells is 20% [256].

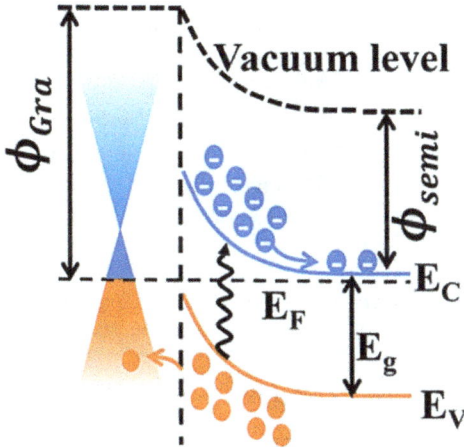

Figure 43   The VB and CB of the MQD-PDES solar cell [256] (with permission to reproduce/reprint from the publisher).

The structure of MQD-PDES solar cell with ZnO and InP QDs is shown in Figure 44(a), which involves Au gate electrode, n-type GaAs substrate, trilayer Gr, Ag contact, ZnO and InP QDs, and Figure 44(b) depicts the I-V diagram [256].

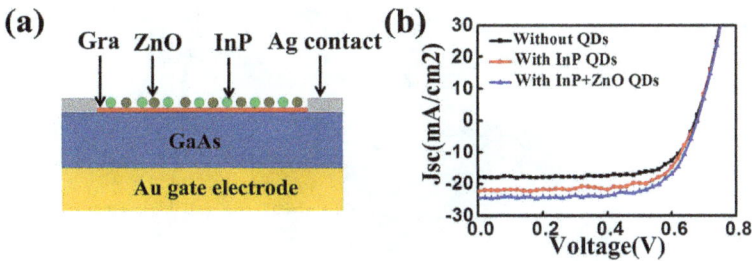

Figure 44   (a) Schema of Gr/GaAs solar cell with two different kinds of semiconductor QDs, (b) The I-V diagram with and without InP and ZnO QDs [256] (with permission to reproduce/reprint from the publisher).

3.21   Microstructure semiconductor based on heavily-doped silicon wafers (MS-HDSW)

The MS-HDSW solar cells is the new technology which have been proposed based on cost-efficient heavily doped Si wafers. In this technology, PCE is created when the diffusion length of J_{ph} is less than penetration depth of the solar radiation. One could achieve PCE from 10.7% up to 15% and also Voc [258-261] for doping concentration as high as $10^{18}/cm^3$ The key parameters in this technology are doping and thickness of the sidewall emitter passivated by SiO_2 and top n+ layer (Table 14) [262]. Figure 45 [262] shows the I-V characteristic of the MS-HDSW solar cells. This technology follows two major goals: 1) UV collector, and 2) highest PCE achievements [262].

Table 14   *MS-HDSW solar cell performance for different n+ layer doping [262] (with permission to reproduce/reprint from the publisher).*

Doping (cm^{-3})	10^{19}	2×10^{19}	3×10^{19}	5×10^{19}	10^{20}
J_{sc} (nA)	6.9	6.9	6.89	6.89	6.85
Voc (V)	0.583	0.583	0.582	0.579	0.563
FF (%)	82.07	82.18	82.22	82.16	81.62
η (%)	14.76	14.77	14.75	14.66	14.06

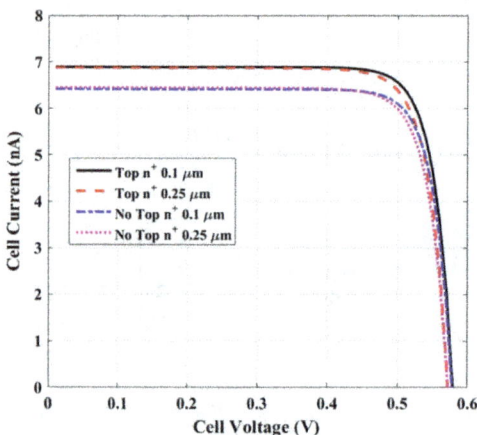

Figure 45   *I-V characteristic of MS-HDSW solar cells for different n+ layer doping [262] (with permission to reproduce/reprint from the publisher).*

The main points in MS-HDSW solar cells for reaching the suitable value is that 1) the thickness must be shallow due to collection of UV CCs and doping must not be very high to inhibit quick collection of UV CCs, For example, both J_{sc} and Voc for thickness of $0.1\mu m$ are much better than $0.25\mu m$, and 2) the doping must not be small to inhibit reduction of FF, as well as J_{sc} and Voc. Figure 46 depicts the MS-HDSW solar cell parameters against the width for two values of emitter doping [262].

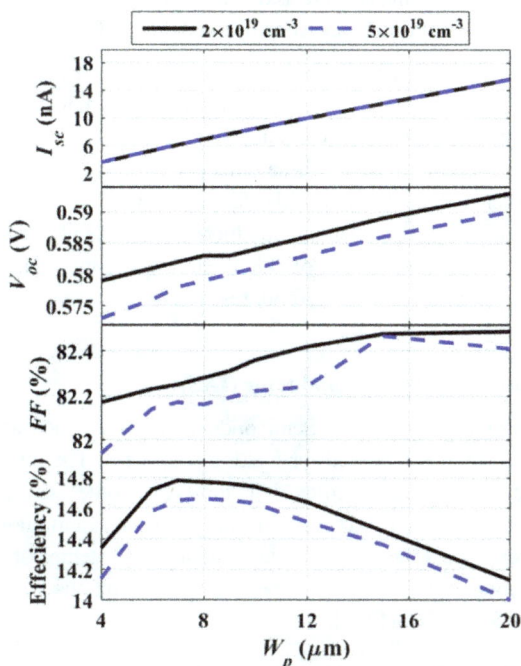

Figure 46   MS-HDSW solar cell parameters against of the width for two values of emitter doping [262] (with permission to reproduce/reprint from the publisher).

3.22   Amorphous semiconductor (AS)

Amorphous semiconductor (AS) solar cells such as a-Si or Amorphous a-InGaZnO4 (a-IGZO) [263] are the solar cells in which their semiconductor structure is non-crystalline, so CIS structure can be used for these kinds of solar cells. The a-IGZO can be utilized in perovskite solar cells as an electron transporting layer (ETL) to increase their PCE [263]. The amorphous (a-Si) was invented in 1975 [3]. To date, different methods such as sputtering, glow discharge, ion plating, electro deposition, CVD or evaporation are used for preparing a-Si. This AS due to having the direct band gap nature, causes a film with about $1\mu m$ thick to receive most of the solar spectrum [3, 264]. Also, hydrogen is commonly used with amorphous semiconductor such a Si and Hydrogen (H), called a-Si:H or Hydrogenated Nanocrystalline Si (nc-Si:H) [52, 113, 202, 265]. One of the advantages of it is that it can be used in thin-film solar cells and n-/p-doped formed by film growth and adding B or P gases [113]. Another AS is ITO/a-Si that is not recommended for solar cells due to having less work function, less control on interfacial layer, and appearance of defects in interface [15, 266].

AS solar cells require p-i-n structures, because the intrinsic layer is the main absorber area with thickness of about $0.4\text{-}0.8\mu m$, but the p areas are very thin about $10nm$. One of the main problems of AS solar cells is the electrical defects due to high illumination and temperature which leads to low resistivity, however these defects can be eliminated by annealing up to 150°C. Another problem is Staebler-Wronski effect which is an intrinsic defect and decrease the PCE by 10-30%, however this phenomenon increases the stability due to more absorbed light in doped area. So, for preventing of this phenomenon, the band gap is widened or narrowed by mixing the germanium or carbon, respectively, so we can have PCE over 10% [52, 202, 267].

3.23   Improved amorphous semiconductor (IAS)

Recently, some Improved Amorphous Semiconductors (IASs) such as Ge-Si alloys and germanium with halogen are employed for making solar cells. These semiconductors have better photoconductive response to solar radiation. Also, other semiconductors such as fluorinated amorphous Si have a good spectral response in contrast to fluorinated material gas which cannot improve the solar cell PCE and require high temperature up to ~300°C. Whereas, the maximum temperature for fluorinated amorphous Si is near 65°C. The maximum PCE of IASs is ~26% [268].

3.24   Field-effect ferroelectric semiconductor (FEFS)

As novel solar cells, field-effect ferroelectric semiconductor (FEFS) solar cells have combined the semiconductor materials with ferroelectric materials. In fact, they are not p-n junction solar cells, but they have high PCE. Figure 47 shows the working principle of FEFS solar cells [46].

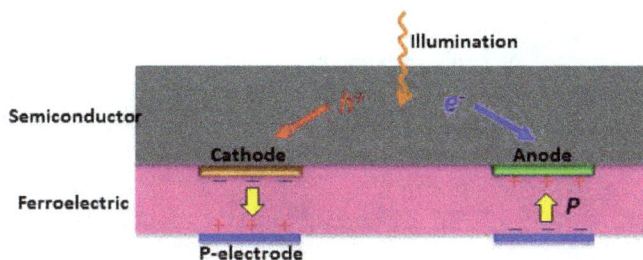

Figure 47   *Working principle of FEFS solar cells [46] (with permission to reproduce/reprint from the publisher).*

For preparing a FEFS solar cell, one usually uses poly-crystalline $BaTiO_3$ and p-type single crystalline Si as the ferroelectric layer and semiconductor absorber, respectively. Figure 48 depicts the I-V curves of a FEFS solar cells under illumination condition [46].

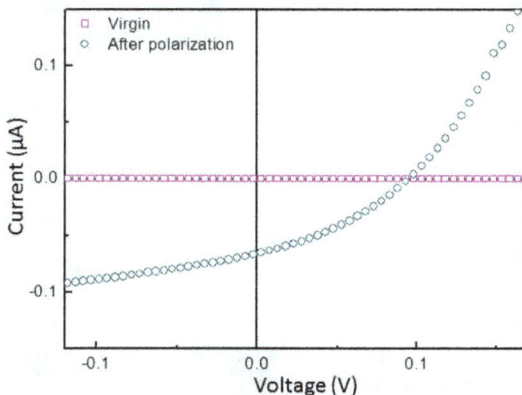

Figure 48   *I-V characteristic of FEFS solar cells [46] (with permission to reproduce/reprint from the publisher).*

There is not any theoretical estimation for maximum PCE of FEFS solar cells due to lack of the thermodynamic limits, but currently the performed experiments have shown increase of PCE from 9% to ~20% in these solar cells [269].

4. Semiconductor modelling in solar cells

All thin film structures of solar cells can be modeled as heterojunction structure with dominant recombination at the interface or as pin diode structure with dominant recombination in the intrinsic layer. In all cases the illuminated I-V characteristics can be represented by a general form as follows (section 2, Eq. 1) [2]:

$$I = I_{ph} - I_s \left(e^{\frac{(V+IR_s)}{nV_T}} - 1 \right) - \frac{V + IR_s}{R_{sh}}$$

This is the general five parameters model whereas clear from the equation and mentioned before, the parameters are I_{ph} is the photocurrent, I_s is the reverse saturation current, n is the ideality factor, R_s is the series resistance and R_{sh} is the shunt resistance.

The ideality factor is sensitive to the conduction mechanism in the solar cell. It is equal to one in case the current is limited by the recombination in the neutral regions around the space charge region of the junction [2, 73]. If the current is dominant by recombination in the space charge region the ideality factor will be equal to 2 [73]. In case of pin diode with dominant recombination in the i-region, the ideality factor will be 2 [68]. In case of recombination at the heterointerface, the ideality factor can take values greater than two [2, 68, 73]. Hence, for Si solar cells, the n value is 1-4 and the larger value is damaged by radiation [67, 270].

R_s is the ohmic series resistance of all neutral layers in the solar cells. The shunt resistance, R_{sh}, is due to partial shorts across the junction mostly due to pinholes and defects in the fabrication process, touching the upper and lower heavily doped layers, or weak junction points.

Good solar cells have low R_s, high R_{sh}, and $n \leq 2$. Also, their I_s must be as small as possible and I_{ph} must be high.

The above model equation (Eq. 1) is valid when the metal electrode interfaces to the active semiconductor regions have low ohmic. In case that these metal semiconductor contacts are rectifying, then they will affect much the illuminated I-V curve reducing its fill factor and changing its shape to an S-shape [271].

Taking this effect into consideration, the equivalent circuit model will be as shown in Figure 49 [271]. The leaky Schottky diode representing the non-ohmic contact normally follows the I-V relation, $I = kV^m$, where k and m are the parameters of the I-V curve that can be experimentally determined [271].

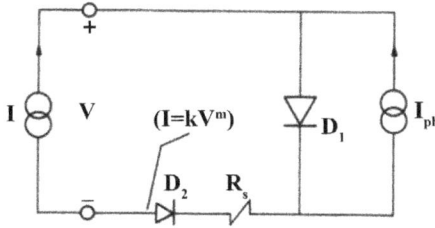

Figure 49   the equivalent circuit model with MS contact [271] (with permission to reproduce/reprint from the publisher).

However, MS contacts in the solar cells must be compromise with low ohmic.

In addition to analytical modelling, there is numerical simulation using solid state device simulators as Silvaco TCAD [2] or COMSOL [272].

In general, for an active material with thickness, δ, of 200nm, we can generate electron hole density or photo current, I_{ph}, of 16mA/cm^2 as follows (Eq. 15):

$$I_{ph} = q\alpha F\delta \tag{15}$$

Where q is the electron charge and equal to $1.619^{-.6}C$, α is the absorption coefficient equal to $510^{14}/cm$, and F is the incident photon flux at AM1 equal to 1017 photons/cm^2.

For analyzing the solar cell models, especially perovskites and OSs, we use two modelling as follows [68]:

1) The analytical modelling can be done as Figure 50.

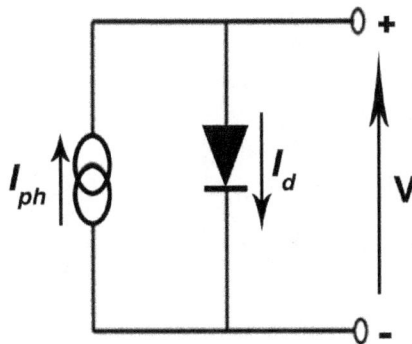

Figure 50   The first analytical modelling.

As you see in Figure 50, the total current is obtained from subtracting the dark diode current from the I_{ph} as follows (Eq. 16):

$$I = I_{ph} - I_d \qquad (16)$$

Assuming that the material absorbs uniformly through its thickness, δ, with the photogeneration rate, G, so, I_{ph} and I_d are calculated as follows (Eqs. 17,18):

$$I_{ph} = qG\delta \qquad (17)$$

$$I_d = I_s(e^{\frac{V}{nV_T}} - 1) \qquad (18)$$

2) The analytical modelling can be done as Figure 51.

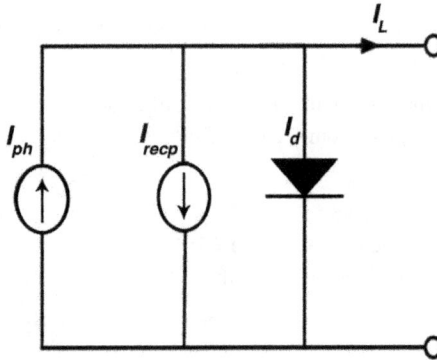

Figure 51   The second analytical modelling.

As you see in Figure 51, the total current is calculated as follows (Eqs. 19,20):

$$I = I_{ph} - I_{recp} - I_d \qquad (19)$$

$$I_{recp} = I_{ph}\frac{\delta}{L_{ef}} \qquad (20)$$

Where I_{recp} is the recombination current and L_{ef} is the effective drift length calculated as follows (Eq. 21):

$$L_{ef} = 3\mu\tau E \qquad (21)$$

Where μ is the mobility in the active layer, τ is the diffusion time or lifetime, and E is the electric field. So, the total current is obtained as follows (Eq. 22):

$$I = I_{ph}(1 - \frac{\delta}{L_{ef}}) - I_d \tag{22}$$

Modeling must be done in accordance with following cases [273-275]:

- Deep bulk levels
- Interface states and band discontinuities
- Optical production
- Steady-state or Direct current (DC) simulations
- Alternative current (AC) simulations
- Exciton transport and dynamics
- Extensions to the basic model
- Numerical solution
- Effect of semiconductor layer width on I-V characteristic
- Effect of semiconductor doping level on I-V characteristic

The three following ways determine the communication between the semiconductor/metal, surface states, semiconductor CB, and VB [131, 203]:

1) Direct passing of electrons from VB to CB which increases with the short circuit current, J_{sco}, when δ is equal to zero. The reverse process is radiative recombination current which is very small.

2) Holes tunnel from semiconductor VB to the metal filled state. In equal current in thermal equilibrium, it moves from metal to the semiconductor. Photo-generated holes reach the semiconductor oxide-interface when the light is happening and influences on the space charge layer. J_{sco} is equal to the number of these carriers arriving at the interface per unit area per unit time. When the relaxation of the carriers close to the interface is quick, an extra carrier (holes) density, δ_p is produced close to the surface. So, the tunneling hole current, J_p is as follow (Eqs. 23,24),

$$J_p = q\delta_p v_p \exp\left(-b_p \chi_p^{1/2} \delta\right) \tag{23}$$

Where

$$v_p = \sqrt{KT/2\pi m_p^*} \tag{24}$$

Where the χ_p is the barrier height for holes, and v_p is the effective thermal velocity of holes in eV. The v_p expression is calculated with the probability for normal incidence on the surface, and Maxwell Boltzmann distribution of velocities. The b_p is also the tunneling constant for holes, K Boltzmann's constant, T absolute temperature, and m_p^* tunneling effective mass for the holes.

3) An electron current, J_n moves from the semiconductor CB to the metal using Richardson-Dushman equation (Eq. 25)

$$J_n = J_{no} \exp\left(-b_n \chi_n^{1/2} \delta\right) [\exp(q\Delta V_D/KT) - 1] \tag{25}$$

Where J_{no} is computed as follows (Eq. 26)

$$J_{no} = q v_n n_o \exp(-qV_D/KT) \tag{26}$$

and v_n is given as follows (Eq. 27)

$$v_n = \sqrt{(KT/2\pi m_n^*)} \tag{27}$$

Where the v_n is the effective thermal velocity of electrons in eV, m_n^* tunneling effective mass for electrons, χ_n is the barrier height for electrons in the CB close to the surface because of the oxide layer, b_n is also tunneling constant for electrons, and J_{no} is the number of electrons existing in the neutral semiconductor CB in the thermal equilibrium. Also V_D is the potential in the space charge layer of the semiconductor. The n_o is the number of electrons existed in the CB of the neutral semiconductor in the thermal equilibrium.

Figure 52 [203] shows the changes in current densities. J_D is current density in surface (depletion) layer [203].

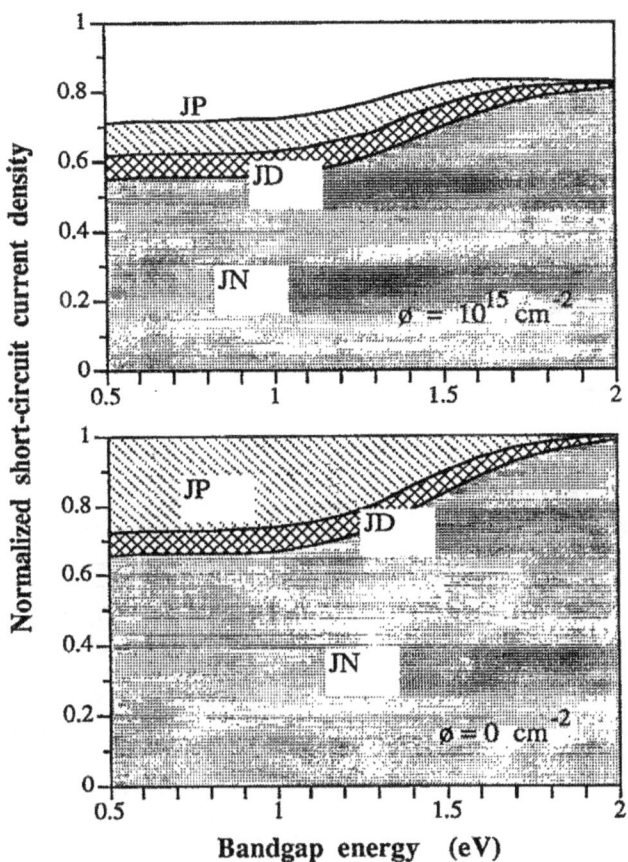

Figure 52   *Changes in current densities [203] (with permission to reproduce/reprint from the publisher).*

Modelling and design of the semiconductor must be measured by nine criteria as below [276-279]:

- Stability
- Durability and insolubility
- Minimizing losses (Figure 53 [279])
- Strong absorption
- Carrier multiplication
- Thermodynamic efficiency limits
- Vastly separated fermi levels
- High exciton mobility
- Easy exciton separation
- Optimal dye hierarchy

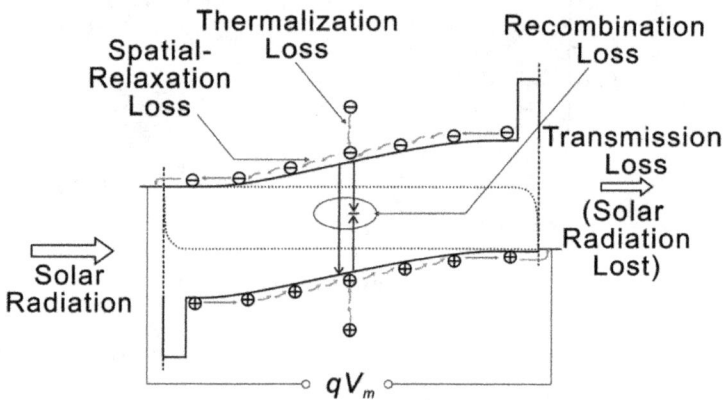

Figure 53   Key losses of a solar cell [279] (with permission to reproduce/reprint from the publisher).

Table 15 compare all the semiconductor solar cells in this book with each other.

Table 15   Comparison of semiconductor solar cells with each other.

Semiconductor solar cell name	η (%)	Disadvantage	Advantage	Ref.
OS	70	Bad electrical performance	1. Light weight 2. Less cost 3. High optical absorption coefficient 4. Low mobility 5. High dissociation energy 6. Less light loss	[67, 68, 77-79]
PS	8.13-12.49		1. Higher absorption coefficient 2. Small energy loss 3. Great stability 4. High electrical conductance	[81, 87]
IOS	1.7-3.13		1. Low cost 2. High efficiency 3. Better electron transporting 4. Better light harvesting 5. Very cost-effective	[42, 65, 89, 98-108]
PAS	3		1. High quantum efficiency 2. Cost-effective	[114]
CPS	6.3×10^{-5}-4.3	1. Poor efficiency 2. Weak FF and J_{sc} 3. Highly dependent on the polymer sample	Linear characteristic	[2, 116-119, 126, 127]
MS/SB	0.72-20.8		1. Highly potential for conversion solar energy to electrical energy 2. Highly minority-carrier collection efficiency 3. Strong current outputs 4. High optical absorption 5. Flexibility with poly crystalline thin films 6. Direct band gap 7. Less temperature 8. Strong radiation resistance 9. Less cost	[86, 128-130]

MOS	> 60	1. Depending on the selection of the solar cell characteristic to m* and E_g 2. Image force effects on Voc and J_{sc} decreases due to barrier height decrease	1. High Voc and J_{sc} 2. High light absorption 3. Less cost	[86, 131, 139]
IMOS	10-> 60		High Voc and J_{sc}	[4]
AMOS	> 60	Dry oxygen is not enough for proper thickness		[86, 132]
SITOS	0.56-1.06		1. Highly compatible with every fluid for substrates 2. Cost-effective 3. Low cost 4. High deposition rate and efficiency	[143, 144]
CIS: 1. MIS 2. SIS 3. EIS		The role of insulating layer in decrease of image force effects	1. Using a p-n junction diode in electronic appliances 2. Low cost 3. Long-term operating lifetime	[3, 134, 145, 146, 150, 151]
MIS	9.5-18.3 [3]	Decrease of PCE due to high temperature and illumination	1. Reduction of shunting 2. Simple fabrication	[3, 170, 173, 174]
SIS	14-16		More stable and efficient than MIS and MS/SB	[33, 179-182]
EIS	4-15 [280]		Using as a photovoltaic output for MIS and SIS	[3, 197]
CIOS	Very low	Very low PCE	More stable and efficient than MIS	[147]
MIOS	Very low	Very low PCE		[198]
MINDS	1-10 [200]	Reduction of Voc and J_{sc} because of high temperature		[199, 200]

MNN-MIS	11	Using nanoscale structure for increasing J_{sc} and optical transmittance	1. Simple fabrication 2. Having the properties of p-n junction solar cells	[201]
PCS	12-34.6	Some of them are not cost-efficient.	1. Most of them are cheap. 2. High radiation resistance 3. High optical absorption	[52, 53, 202, 204, 210, 211]
MicoSem	12-15	Not proper for 540 < temp < 110°K	1. Having many commercial applications 2. Less cost	[213, 215]
NS	5.3-≥ 18	Control of size and mixture of the nanostructure due to high optical absorption	1. Good alternative for poly Si semiconductors 2. They are not nonuniform. 3. Less cost 4. Simple fabrication	[49, 109, 216-219]
QD-IBS	63.2		High Voc, J_{sc} and PCE	[227, 245-248]
MQD-PDES	8.57-20		Great stability	[256]
MS-HDSW	10.7-15	Doping must not be very high or very small	1. Cost-effective 2. High Voc 3. High UV collection	[262]
AS	> 10	1. Some of them have less work function, less control and interface defects 2. Electrical defects due to high illumination and temperature	1. High optical absorption 2. It can be used in thin-film solar cells.	[15, 52, 113, 202, 266, 267]
IAS	~ 20		Better photoconductive response to solar radiation	[268]
FEFS	9-~ 20	There is not a theoretical estimation for maximum PCE due to lack of thermodynamic limits.		[269]

Conclusion and Future work

This book shows that the solar cells structures and technology have been greatly advanced such that one enhanced the photo conversion efficiency appreciably. At the same time, the production cost has been greatly reduced. There are some promising solar cell structures such as the MIS and SIS structures. Also, hybrid perovskites proved to be a strong candidate for low-cost solar cells except they suffer from instability. A new type of solar cell which is the heavily doped Si solar cell must receive more research as it is a strong candidate for low-cost solar cell production as the heavily doped Si is relatively cheap.

In the future, lower band gap particulate semiconductors, e.g., cadmium sulfide and lead oxide must be studied more. Also, decrease of the overall reflectance will need dielectric coatings whose refractive index is near to 2.35 – i.e., for 60Å Au on GaAs – or sufficient multiple layer AR coatings. The angle variability of the sunlight incident must be identified for the design of the practical AR coatings.

Today, thin-insulator growth methods are continually being enhanced with other information about thick-insulator MOS structures and can result in the fabrication technique of n-ITO-SiO_2-p-Si solar cells with high Voc and also PCE [196].

For calculation of the MIS solar cells different interfacial film parameters and band gap are needed. As a regard, three models must be kept in view according to the recombination kinetics of interface state: Model 1) interface states being in equilibrium with the metal, Model 2) interface states being in equilibrium with majority of the semiconductor carriers, and Model 3) interface states being in equilibrium with the minority of the semiconductor carriers. In all these three models, SB cells and the PCE of low barrier height increase very much by the addition of an insulator layer [281].

Also, the future, optimized n-p-n solar cell structures can be used in a way in which the illumination performance will enhance because of the removing of the structure notches.

References

1. Zekry, A., *A road map for transformation from conventional to photovoltaic energy generation and its challenges.* Journal of King Saud University - Engineering Sciences, 2020. **32**(7): p. 407-410.
2. Zekry, A., A. Shaker, and M. Salem, *Chapter 1 - Solar Cells and Arrays: Principles, Analysis, and Design*, in *Advances in Renewable Energies and Power Technologies*, I. Yahyaoui, Editor. 2018, Elsevier. p. 3-56.
3. Singh, R., M.A. Green, and K. Rajkanan, *Review of conductor-insulator-semiconductor (CIS) solar cells.* Solar Cells, 1981. **3**(2): p. 95-148.
4. Sircar, P. and R. Dat, *Improved GaAs metal-oxide semiconductor solar cells with a spin-on oxide layer.* Canadian Journal of Physics, 1981. **59**(5): p. 716-717.

5.   DuBow, J.B., D.E. Burk, and J.R. Sites. *Solar cells of indium tin oxide on silicon.* in *1975 International Electron Devices Meeting.* 1975.

6.   Franz, S.L., et al. *A comparison of In2O3/Si and SnO2/Si heterojunction solar cells with polycrystalline and with monocrystalline substrates.* in *Proc. Nat. Workshop on low cost polycrystalline silicon solar cells.* 1976. Dallas, TX.

7.   DuBow, J.B., D.E. Burk, and J.R. Sites, *Efficient photovoltaic heterojunctions of indium tin oxides on silicon.* Applied Physics Letters, 1976. **29**(8): p. 494-496.

8.   Lai, S.W., et al. *In2O3/Si heterojunction solar cells.* in *11th IEEE Photovoltaic specialists conference* 1975.

9.   Wang, E.Y. and R.N. Legge, *General properties of SnO2-GaAs and SnO2-Ge heterojunction photovoltaic cells.* IEEE Transactions on Electron Devices, 1978. **25**(7): p. 800-803.

10. Burk, D.E., J.B. DuBow, and J.R. Sites. *Fabrication of OSOS cells by neutral ion beam sputtering.* in *12th IEEE Photovoltaic Specialists conference* 1976.

11. Nash, T.R. and R.L. Anderson. *Degradation of SnO2/Si heterojunction solar cells.* in *12th IEAEE Photovoltaic specialists conference* 1976.

12. Franz, S., G. Kent, and R.L. Anderson, *Heterojunction solar cells of SnO2/Si.* Journal of Electronic Materials, 1977. **6**(2): p. 107-123.

13. Sree Harsha, K.S., et al., *n-indium tin oxide/p-indium phosphide solar cells.* Applied Physics Letters, 1977. **30**(12): p. 645-646.

14. Manifacier, J.C. and L. Szepessy, *Efficient sprayed In2O3 : Sn n-type silicon heterojunction solar cell.* Applied Physics Letters, 1977. **31**(7): p. 459-462.

15. Mizrah, T. and D. Adler, *Indium—Tin—Oxide—Silicon heterojunction photovoltaic devices.* IEEE Transactions on Electron Devices, 1977. **24**(4): p. 458-462.

16. Thompson, W.G., et al., *Intensity effects in SnO2—Si heterojunction solar cells.* IEEE Transactions on Electron Devices, 1977. **24**(4): p. 463-467.

17. Nash, T.R. and R.L. Anderson, *Accelerated life tests of SnO2—Si heterojunction solar cells.* IEEE Transactions on Electron Devices, 1977. **24**(4): p. 468-472.

18. Chang, N.S. and J.R. Sites, *Electronic characterization of indium tin oxide/silicon photodiodes.* Journal of Applied Physics, 1978. **49**(9): p. 4833-4837.

19. Wang, E.Y. and L. Hsu, *ChemInform Abstract: DETERMINATION OF ELECTRON AFFINITY OF INDIUM OXIDE (IN2O3) FROM ITS HETEROJUNCTION PHOTOVOLTAIC PROPERTIES.* Chemischer Informationsdienst, 1978. **9**(51).

20. Thompson, W.G. and R.L. Anderson, *Electrical and photovoltaic characteristics of indium-tin oxide/silicon heterojunctions.* Solid-State Electronics, 1978. **21**(4): p. 603-608.

21. Singh, R. and J. Shewchun, *The importance of the electron affinity of oxide-semiconductors as used in solar cells.* Appl. Phys. Lett. , 1978. **33**: p. 601-603.

22. Cheek, G., et al., *Fabrication and characterization of indium tin oxide (ITO)/polycrystalline silicon solar cells.* Applied Physics Letters, 1978. **33**(7): p. 643-645.

23. Ghosh, A.K., C. Fishman, and T. Feng, *SnO2/Si solar cells—heterostructure or Schottky-barrier or MIS-type device.* Journal of Applied Physics, 1978. **49**(6): p. 3490-3498.

24. Fishman, C., T. Feng, and A.K. Ghosh. *Life testing and stability of SnO2/n-Si solar cells.* in *1978 International Electron Devices Meeting.* 1978.

25. Manasevit, H.M., et al. *The properties of homoepitixal InP films prepared by the MO-CVD process for the fabrication of heterojunction solar cells.* in *13th IEEE Photovoltaic Specialists Conference* 1978.

26. Fahrenbruch, A.L., et al. *Recent investigations of metal oxide/CdTe heterojunction solar cells.* in *13thIEEE Photovoltaic Specialists Conf.* 1978.

27. Feng, T., C. Fishman, and A.K. Ghosh. *Barrier heights and inter facial effects in SnO2/Si solar cells.* in *13th IEEE Photovoltaic specialists conf.* . 1978.

28. Bachmann, K.J., et al. *Indium-tin oxide/indium phosphide and indium-tin oxide/gallium arsenide solar cells.* in *13th IEEE Photovoltaic Specialists Conf.* 1978.

29. Shewchun, J., et al. *The photovoltaic effect in interfacial layer heterojunctions or semiconductor-insulator-semiconductor diodes: indium-tin-oxide on silicon, gallium arsenide and indium phosphide.* in *13th IEEE Photovoltaic Specialists Conf.* . 1978.

30. Hsu, L. and E.Y. Wang. *Photovoltaic properties of In2O3/semiconductor heterojunction solar cells.* in *13th IEEE Photovoltaic Specialists Conf.* 1978.

31. Kazmerski, L.L. and P. Sheldon. *Fabrication and characterization of ITO/CuInSe2 photovoltaic heterojunctions.* in *13th IEEE Photovoltaic Specialists Conf.* 1978.

32. Feng, T., A.K. Ghosh, and C. Fishman, *Angle-of-incidence effects in electron-beam-deposited SnO2/Si solar cells.* Applied Physics Letters, 1979. **34**(3): p. 198-199.

33. Shewchun, J., et al., *The operation of the semiconductor-insulator-semiconductor solar cell: Experiment.* Journal of Applied Physics, 1979. **50**(4): p. 2832-2839.

34. Bachmann, K.J., et al., *Solar-cell characteristics and interfacial chemistry of indium-tin-oxide/indium phosphide and indium-tin-oxide/gallium arsenide junctions.* Journal of Applied Physics, 1979. **50**(5): p. 3441-3446.

35. Winn, O.H., S.L. Franz, and R.L. Anderson, *Static optoelectronic characteristics of SnO2/V2O5 : P2O5/Si heterojunctions.* Journal of Applied Physics, 1979. **50**(5): p. 3758-3761.

36. Singh, R., *Discussion of "Determination of Electron Affinity of In2 O 3 from Its Heterojunction Photovoltatic Properties" [E. Y. Wang, and L. Hsu (pp. 1328–1331, Vol. 125, No. 8)].* Journal of The Electrochemical Society, 1979. **126**(6): p. 1081-1081.

37. Fishman, C., A.K. Ghosh, and T. Feng, *Stability of SnO2/n-Si solar cells.* Solar Energy Materials, 1979. **1**(1): p. 181-185.

38. Feng, T., A.K. Ghosh, and C. Fishman, *Efficient electron-beam-deposited ITO/n-Si solar cells.* Journal of Applied Physics, 1979. **50**(7): p. 4972-4974.

39. Osterwald, C., et al., *Molybdenum trioxide (MoO3)/silicon photodiodes.* Applied Physics Letters, 1979. **35**(10): p. 775-776.

40. Singh, R., et al., *Optimization of oxide-semiconductor/base-semiconductor solar cells.* IEEE Transactions on Electron Devices, 1980. **27**(4): p. 656-662.
41. Hezel, R., *Solar cells composed of semiconductor materials.* 1981.
42. Xu, T. and Q. Qiao, *Conjugated polymer–inorganic semiconductor hybrid solar cells.* Energy & Environmental Science, 2011. **4**(8): p. 2700-2720.
43. Hussein, H. and A. Yazdani, *Doping the bismuth into the host's Cu2ZnSnS4 semiconductor as a novel material for thin film solar cell.* Results in Physics, 2019. **12**: p. 1586-1595.
44. I.S. Amiri, M.A., *Development of Solar Cell Photovoltaics: Introduction and Working Principles*, in *Introducing CTS (Copper-Tin-Sulphide) as a Solar Cell by Using Solar Cell Capacitance Simulator (SCAPS).* 2019, Springer, Cham. p. 1-14.
45. Suemasu, T.T., *Semiconductor material, solar cell using the semiconductor material, and methods for producing the semiconductor material and the solar cell.* 2014.
46. Wang, W., et al. *Field-effect ferroelectric-semiconductor solar cells.* in *2014 IEEE 40th Photovoltaic Specialist Conference (PVSC).* 2014.
47. Peter, I.J., et al. *Performance of TiO2/CdS/Bi2S3 heterostructure based semiconductor sensitized solar cell.* in *AIP Conference Proceedings.* 2019.
48. Schock, H.W. *Thin Film Compound Semiconductor Solar Cells: An Option for Large Scale Applications?* in *Tenth E.C. Photovoltaic Solar Energy Conference.* 1991. Dordrecht: Springer Netherlands.
49. Das, B., S.P. McGinnis, and P. Sines, *High-efficiency solar cells based on semiconductor nanostructures.* Solar Energy Materials and Solar Cells, 2000. **63**(2): p. 117-123.
50. Zweibel, K. *Thin film technologies project FY96 Summary.* in *Technology's Critical Role in Energy and Environmental Markets conference.* 1998. Albuquerque, New Mexico.
51. Kay, A. and M. Graetzel, *Artificial photosynthesis. 1. Photosensitization of titania solar cells with chlorophyll derivatives and related natural porphyrins.* The Journal of Physical Chemistry, 1993. **97**(23): p. 6272-6277.
52. Möller, H.J., *New Materials: Semiconductors for Solar Cells.* Materials Science and Technology, 2013.
53. Islam, M.O., et al. *III-nitride compound semiconductors for solar cell.* in *2008 2nd National Workshop on Advanced Optoelectronic Materials and Devices.* 2008.
54. Keheze, F.M., M.P. Karimi, and N. Walter, *Copper based solar cell materials.* 2013: LAP LAMBERT Academic Publishing.
55. *Photovoltaic Technologies*, in *NREL Report.*
56. Litvinov, V.G., et al. *Measuring complex for analysis of recombination deep traps in semiconductor solar cells.* in *2015 IEEE International Conference on Industrial Technology (ICIT).* 2015.
57. Genis, A.P., et al., *Efficient indium tin oxide/polycrystalline silicon semiconductor-insulator-semiconductor solar cells.* Appl. Phys. Lett., 1980. **37**: p. 77-79.

58. Sahbel, A., et al., *Experimental Performance Characterization of Photovoltaic Modules Using DAQ.* Energy Procedia, 2013. **36**: p. 323-332.

59. Sze, S.M. and K.K. Ng, *Physics of Semiconductor Devices.* 2006: John Wiley & Sons.

60. Chen, C.J., *Semiconductor Solar Cells.* Physics of Solar Energy, 2011: p. 177-210.

61. Yang, W., et al., *Template-electrodeposition preparation and structural properties of CdS nanowire arrays.* Microelectronic Engineering, 2006. **83**(10): p. 1971-1974.

62. Chatterjee, S., U. Dasgupta, and A.J. Pal. *Solar Cells: Materials Beyond Silicon.* in *Energy Engineering.* 2017. Singapore: Springer Singapore.

63. Kaneko, K., *Recent Advancement of Semiconductor Materials and Devices I: Physical Properties of Oxide Semiconductors and Their Applications for Electrical Devices.* Journal of the Society of Materials Science, Japan, 2017. **66**(1): p. 58-65.

64. Stroyuk, O., *Semiconductor-Based Liquid-Junction Photoelectrochemical Solar Cells*, in *Solar Light Harvesting with Nanocrystalline Semiconductors*, O. Stroyuk, Editor. 2018, Springer International Publishing: Cham. p. 161-240.

65. Zhuiykov, S., *Chapter 7 - Nanostructured Semiconductor Composites for Solar Cells*, in *Nanostructured Semiconductors (Second Edition)*, S. Zhuiykov, Editor. 2018, Woodhead Publishing. p. 353-412.

66. Avrutin, V., N. Izyumskaya, and H. Morkoç, *Semiconductor solar cells: Recent progress in terrestrial applications.* Superlattices and Microstructures, 2011. **49**(4): p. 337-364.

67. Fang, P.H., *Analysis of conversion efficiency of organic-semiconductor solar cells.* J. Appl. Phys. , 1974. **45**: p. 4672-4673.

68. Zekry, A., I. Yahyaoui, and F. Tadeo. *Generic Analytical Models for Organic and Perovskite Solar Cells.* in *2019 10th International Renewable Energy Congress (IREC).* 2019.

69. Sulaiman, K., et al., *Organic Semiconductors: Applications in Solar Photovoltaic and Sensor Devices.* Materials Science Forum, 2013. **737**: p. 126-132.

70. Fujiseki, T., et al., *Organic Semiconductors*, in *Spectroscopic Ellipsometry for Photovoltaics: Volume 2: Applications and Optical Data of Solar Cell Materials*, H. Fujiwara and R.W. Collins, Editors. 2018, Springer International Publishing: Cham. p. 427-469.

71. Simon, J., J.J. André, and E. Conwell, *Molecular Semiconductors: Photoelectrical Properties and Solar Cells.* Physics Today, 1986. **39**(2): p. 75-76.

72. Maslenikov, S.V. and M.I. Fedorov, *Organic semiconductor solar cells with a heterojunction.* Russian Physics Journal, 1997. **40**(1): p. 60-63.

73. Abdel Naby, M., et al., *Dependence of dark current on zinc concentration in ZnxCd1−xS/ZnTe heterojunctions.* Solar Energy Materials and Solar Cells, 1993. **29**(2): p. 97-108.

74. Abdelaziz, W., et al., *Possible efficiency boosting of non-fullerene acceptor solar cell using device simulation.* Optical Materials, 2019. **91**: p. 239-245.

75. Jin, R., et al., *Design of perylene-diimides-based small-molecules semiconductors for organic solar cells.* Molecular Physics, 2017. **115**(14): p. 1591-1597.

76. Tu, Q., et al., *Ladder-type heteroheptacene-cored semiconductors for small-molecule solar cells.* Dyes and Pigments, 2018. **149**: p. 747-754.

77. Jonathan, B.W., et al., *Investigations of materials and device structures for organic semiconductor solar cells.* Optical Engineering, 1993. **32**(8): p. 1921-1934.

78. Neal, R.A., et al. *Critical interfaces in new solar cell materials: organic heterojunctions and heterojunctions involving semiconductor nanoparticles.* in *Proc.SPIE.* 2008.

79. Guo, H., et al., *Effect of ITO film deposition conditions on ITO and CdS films of semiconductor solar cells.* Optik, 2017. **140**: p. 322-330.

80. El-Khozondar, H.J., et al., *Two layers corrugated semiconductor solar cell.* Optik, 2019. **181**: p. 933-940.

81. Izquierdo, M.A., R. Broer, and R.W.A. Havenith, *Theoretical Study of the Charge Transfer Exciton Binding Energy in Semiconductor Materials for Polymer:Fullerene-Based Bulk Heterojunction Solar Cells.* The Journal of Physical Chemistry A, 2019. **123**(6): p. 1233-1242.

82. Cohen, M.J. and J.S. Harris, *WA-B3 polymer-semiconductor solar cells.* IEEE Transactions on Electron Devices, 1978. **25**(11): p. 1355-1355.

83. Chiang, C.K., et al., *Electrical conductivity of (SN)x.* Solid State Communications, 1976. **18**(11): p. 1451-1455.

84. Bright, A.A., et al., *Optical Reflectance of Polymeric Sulfur Nitride Films from the Ultraviolet to the Infrared.* Physical Review Letters, 1975. **34**(4): p. 206-209.

85. Scranton, R.A., *The Schottky barriers produced by polymeric sulfur nitride on compound semiconductors.* J. Appl. Phys. , 1977. **48**: p. 3838-3842.

86. Stirn, R.J. and Y.C.M. Yeh, *Technology of GaAs metal—Oxide—Semiconductor solar cells.* IEEE Transactions on Electron Devices, 1977. **24**(4): p. 476-483.

87. Shi, S., et al., *A Narrow-Bandgap n-Type Polymer Semiconductor Enabling Efficient All-Polymer Solar Cells.* Advanced Materials, 2019. **31**(46): p. 1905161.

88. Abdelaziz, W., et al., *Numerical study of organic graded bulk heterojunction solar cell using SCAPS simulation.* Solar Energy, 2020. **211**: p. 375-382.

89. Yi, J., J.H. Yun, and H.K. Kim, *Progress in Low Cost and High Efficiency Silicon and Inorganic Semiconductor Solar Cells.* Israel Journal of Chemistry, 2015. **55**(10): p. 1049-1049.

90. Nakane, A., et al., *Inorganic Semiconductors and Passivation Layers*, in *Spectroscopic Ellipsometry for Photovoltaics: Volume 2: Applications and Optical Data of Solar Cell Materials*, H. Fujiwara and R.W. Collins, Editors. 2018, Springer International Publishing: Cham. p. 319-426.

91. Zhang, G., et al., *Semiconductor nanostructure-based photovoltaic solar cells.* Nanoscale, 2011. **3**(6): p. 2430-2443.

92. Ye, C., et al., *Inorganic semiconductor nanoarrays as photoanodes for solar cells*, in *Handbook of innovative nanomaterials: From Syntheses to Applications* L.W. Xiaosheng Fang, Editor. 2012.

93.   Zheng, M.J., et al., *Fabrication and optical properties of large-scale uniform zinc oxide nanowire arrays by one-step electrochemical deposition technique.* Chemical Physics Letters, 2002. **363**(1): p. 123-128.

94.   Phok, S., S. Rajaputra, and V.P. Singh, *Copper indium diselenide nanowire arrays by electrodeposition in porous alumina templates.* Nanotechnology, 2007. **18**(47): p. 475601.

95.   Takahashi, K., et al., *Synthesis and Electrochemical Properties of Single-Crystal V2O5 Nanorod Arrays by Template-Based Electrodeposition.* The Journal of Physical Chemistry B, 2004. **108**(28): p. 9795-9800.

96.   Lee, J., et al., *Tuning the crystallinity of thermoelectric Bi2Te3nanowire arrays grown by pulsed electrodeposition.* Nanotechnology, 2008. **19**(36): p. 365701.

97.   Ho, Y.-R. and M.-W. Lee, *AgSbS2 semiconductor-sensitized solar cells.* Electrochemistry Communications, 2013. **26**: p. 48-51.

98.   Huynh, W.U., X. Peng, and A.P. Alivisatos, *CdSe Nanocrystal Rods/Poly(3-hexylthiophene) Composite Photovoltaic Devices.* Advanced Materials, 1999. **11**(11): p. 923-927.

99.   Huynh, W.U., J.J. Dittmer, and A.P. Alivisatos, *Hybrid Nanorod-Polymer Solar Cells.* Science, 2002. **295**(5564): p. 2425.

100. Gur, I., et al., *Air-Stable All-Inorganic Nanocrystal Solar Cells Processed from Solution.* Science, 2005. **310**(5747): p. 462.

101. Gur, I., et al., *Hybrid Solar Cells with Prescribed Nanoscale Morphologies Based on Hyperbranched Semiconductor Nanocrystals.* Nano Letters, 2007. **7**(2): p. 409-414.

102. Milliron, D.J., M. Gur, and A. Paul Alivisatos, *Hybrid Organic-Nanocrystal Solar Cells.* MRS Bulletin, 2005. **30**(1): p. 41-44.

103. Bouclé, J., P. Ravirajan, and J. Nelson, *Hybrid polymer–metal oxide thin films for photovoltaic applications.* Journal of Materials Chemistry, 2007. **17**(30): p. 3141-3153.

104. Bouclé, J., et al., *Hybrid Solar Cells from a Blend of Poly(3-hexylthiophene) and Ligand-Capped TiO2 Nanorods.* Advanced Functional Materials, 2008. **18**(4): p. 622-633.

105. Luther, J.M., et al., *Structural, Optical, and Electrical Properties of Self-Assembled Films of PbSe Nanocrystals Treated with 1,2-Ethanedithiol.* ACS Nano, 2008. **2**(2): p. 271-280.

106. Yu, P., et al., *Nanocrystalline TiO2 Solar Cells Sensitized with InAs Quantum Dots.* The Journal of Physical Chemistry B, 2006. **110**(50): p. 25451-25454.

107. Bansal, N., et al., *Solution Processed Polymer–Inorganic Semiconductor Solar Cells Employing Sb2S3 as a Light Harvesting and Electron Transporting Material.* Advanced Energy Materials, 2013. **3**(8): p. 986-990.

108. Zhuiyko, S., *6 - Nanostructured semiconductor composites for solar cells*, in *Nanostructured Semiconductor Oxides for the Next Generation of Electronics and Functional Devices*, S. Zhuiykov, Editor. 2014, Woodhead Publishing. p. 267-320.

109. Sun, K., et al., *Compound Semiconductor Nanowire Solar Cells.* IEEE Journal of Selected Topics in Quantum Electronics, 2011. **17**(4): p. 1033-1049.

110. Li, Y., et al., *BiVO4 semiconductor sensitized solar cells.* Science China Chemistry, 2015. **58**(9): p. 1489-1493.

111. Taranekar, P., et al., *Hyperbranched Conjugated Polyelectrolyte Bilayers for Solar-Cell Applications.* Journal of the American Chemical Society, 2007. **129**(29): p. 8958-8959.

112. Qiao, Q., Y. Xie, and J.J.T. McLeskey, *Organic/Inorganic Polymer Solar Cells Using a Buffer Layer from All-Water-Solution Processing.* The Journal of Physical Chemistry C, 2008. **112**(26): p. 9912-9916.

113. Jiang, C.-S., *Microscopic Electrical Characterization of Inorganic Semiconductor-Based Solar Cell Materials and Devices Using AFM-Based Techniques,* in *Scanning Probe Microscopy in Nanoscience and Nanotechnology 2,* B. Bhushan, Editor. 2011, Springer Berlin Heidelberg: Berlin, Heidelberg. p. 723-790.

114. Weisberg, L.R., C.F. Grain, and R.R. Addiss, *Particulate semiconductor solar cells.* Applied Physics Letters, 1975. **27**(8): p. 440-441.

115. Kanicki, J. *A.C.S. polymer preprints.* in *Proceedings of the Symposium on Conducting Polymers at the American Chemical Meeting.* 1982. Las Vegas, U.S.A.

116. Donckt, E.V., et al., *Photovoltaic properties of the poly-2-vinylpyridine iodine complex—SnO2 system.* Journal of Applied Polymer Science, 1982. **27**(1): p. 1-9.

117. Kanicki, J., et al. *Organic Photovoltaic Materials: Polyacetylene.* in *Fourth E.C. Photovoltaic Solar Energy Conference.* 1982. Dordrecht: Springer Netherlands.

118. Tsukamoto, J., et al., *Characteristics of Schottky barrier solar cells using polyacetylene, (CH)x.* Synthetic Metals, 1982. **4**(3): p. 177-186.

119. Weinberger, B.R., M. Akhtar, and S.C. Gau, *Polyacetylene photovoltaic devices.* Synthetic Metals, 1982. **4**(3): p. 187-197.

120. Cadene, M., et al. *Cadmium Sulfide Polyacetylene Photovoltaic Hererojunction.* in *Fourth E.C. Photovoltaic Solar Energy Conference.* 1982. Dordrecht: Springer Netherlands.

121. Konagai, M., et al., *Electrical Properties of Polyacetylene Thin Films and Application to Solar Cells.* Journal of the Institute of Electrical Engineers of Japan. A, 1981. **101**(2): p. 103-108.

122. Ozaki, M., et al., *Semiconductor properties of polyacetylene p-(CH)x : n-CdS heterojunctions.* Journal of Applied Physics, 1980. **51**(8): p. 4252-4256.

123. Inganäs, O., T. Skotheim, and I. Lundström, *Schottky Barrier Formation Between Polypyrrole and Crystalline and Amorphous Hydrogenated Silicon.* Physica Scripta, 1982. **25**(6A): p. 863-867.

124. Chen, S.N., et al., *Polyacetylene, (CH)x: Photoelectrochemical solar cell.* Applied Physics Letters, 1980. **36**(1): p. 96-98.

125. Shirakawa, H., et al., *Polyacetylene film: A new electrode material for photoenergy conversion.* Synthetic Metals, 1981. **4**(1): p. 43-49.

126. Kanicki, J., *REVIEW OF CONDUCTOR-POLYMERIC SEMICONDUCTOR SOLAR CELLS.* J. Phys. Colloques, 1983. **44**(C3): p. C3-529-C3-535.

127. Kalachyova, Y., et al., *Flexible Conductive Polymer Film Grafted with Azo-Moieties and Patterned by Light Illumination with Anisotropic Conductivity.* Polymers, 2019. **11**(11).

128. Tirrito, E. *Design, fabrication and characterization of metal-semiconductor solar cells.* 2011.

129. Yeh, Y.C.M., F.P. Ernest, and R.J. Stirn, *Practical antireflection coatings for metal-semiconductor solar cells.* Journal of Applied Physics, 1976. **47**(9): p. 4107-4112.

130. Shewchun, J., R. Singh, and M.A. Green, *Theory of metal-insulator-semiconductor solar cells.* Journal of Applied Physics, 1977. **48**(2): p. 765-770.

131. Srivastava, G.P., P.K. Bhatnagar, and S.R. Dhariwal, *Theory of metal-oxide-semiconductor solar cells.* Solid-State Electronics, 1979. **22**(6): p. 581-587.

132. Stirn, R.J. and Y.C.M. Yeh, *A 15% efficient antireflection-coated metal-oxide-semiconductor solar cell.* Applied Physics Letters, 1975. **27**(2): p. 95-98.

133. Fabre, E., *MIS silicon solar cells.* Applied Physics Letters, 1976. **29**(9): p. 607-610.

134. Green, M.A. and R.B. Godfrey, *MIS solar cell—General theory and new experimental results for silicon.* Applied Physics Letters, 1976. **29**(9): p. 610-612.

135. Anderson, W.A., J.K. Kim, and A.E. Delahoy, *Barrier height modification in silicon Schottky (MIS) solar cells.* IEEE Transactions on Electron Devices, 1977. **24**(4): p. 453-457.

136. Kar, S. and R. Varghese, *On the mechanism of degradation in Si/SiOx/Ag metal oxide semiconductor solar cells.* Journal of Applied Physics, 1982. **53**(6): p. 4435-4440.

137. Krawczyk, S.K., A. Jakubowski, and M. Żurawska, *Temperature dependence of the short-circuit current in MIS solar cells.* Solar Cells, 1981. **4**(2): p. 187-194.

138. Bhatnagar, P.K., et al., *Temperature considerations in metal-oxide-semiconductor solar cells.* Solar Cells, 1984. **11**(3): p. 263-267.

139. Ojha, V.N., et al., *Image force effects in metal-oxide-semiconductor solar cells.* Journal of Applied Physics, 1982. **53**(3): p. 1734-1736.

140. Wang, E.Y. and K.A. Pandelišev, *The effect of chemical surface treatments on non-native (Bi2O3) GaAs metal-insulator-semiconductor solar cells.* Journal of Applied Physics, 1981. **52**(7): p. 4818-4820.

141. Brinker, D.J. and E.Y. Wang, *GaAs MIS solar cells with evaporated tin oxide interfacial layers.* IEEE Transactions on Electron Devices, 1981. **28**(9): p. 1097-1098.

142. Wang, B., et al., *Solution-Processed All-Oxide Transparent High-Performance Transistors Fabricated by Spray-Combustion Synthesis.* Advanced Electronic Materials, 2016. **2**(4): p. 1500427.

143. Manifacier, J.C., et al. *Some Comments On Sprayed Ito/Semiconductor Solar Cells.* in *Fourth E.C. Photovoltaic Solar Energy Conference.* 1982. Dordrecht: Springer Netherlands.

144. Mohammad, T., V. Kumar, and V. Dutta, *Spray deposited indium doped tin oxide thin films for organic solar cell application.* Physica E: Low-dimensional Systems and Nanostructures, 2020. **117**: p. 113793.

145. Green, M.A., *Effects of pinholes, oxide traps, and surface states on MIS solar cells.* Applied Physics Letters, 1978. **33**(2): p. 178-180.

146. Singh, R., *Theory of Metal-Insulator-Semiconductor (MIS) and Semiconductor-Insulator-Semiconductor (SIS) solar cells.* 1979, McMaster University.

147. Loutfy, R.O., Y.-H. Shing, and D. Krish Murti, *Conductor-insulator-semiconductor organic solar cells.* Solar Cells, 1982. **5**(4): p. 331-341.

148. Bhattacharya, K., P. Basu, and H. Saha, *Back wall schottky barrier solar cell with an interfacial layer.* physica status solidi (a), 1977. **41**(1): p. 317-321.

149. Green, M.A., *The depletion layer collection efficiency for p-n junction, Schottky diode, and surface insulator solar cells.* Journal of Applied Physics, 1976. **47**(2): p. 547-554.

150. Green, A.M., *Properties and Applications of the Metal-Insulator-Semiconductor (MIS) Tunnel Diode.* 1974, McMaster University.

151. Green, M.A., F.D. King, and J. Shewchun, *Minority carrier MIS tunnel diodes and their application to electron- and photo-voltaic energy conversion—I. Theory.* Solid-State Electronics, 1974. **17**(6): p. 551-561.

152. Singh, R. and J. Shewchun, *Tunnel MIS solar cells.* Journal of Vacuum Science and Technology, 1977. **14**(1): p. 89-91.

153. Anderson, W.A., et al. *Variables which influence silicon Schottky solar cell performance.* in *1975 International Electron Devices Meeting.* 1975.

154. Charlson, E.J. and J.C. Lien, *An Al p-silicon MOS photovoltaic cell.* Journal of Applied Physics, 1975. **46**(9): p. 3982-3987.

155. Shevenock, S., S. Fonash, and J. Geneczko. *Studies of M-I-S type solar cells fabricated on silicon.* in *1975 International Electron Devices Meeting.* 1975.

156. Peckerar, M., H.C. Lin, and R.L. Kocher. *Open circuit voltage of MIS Schottky diode solar cells.* in *1975 International Electron Devices Meeting.* 1975.

157. Lillington, D.R. and W.G. Townsend, *Effects of interfacial oxide layers on the performance of silicon Schottky-barrier solar cells.* Applied Physics Letters, 1976. **28**(2): p. 97-98.

158. Kipperman, A.H.M. and M.H. Omar, *Improved efficiency of MIS-silicon solar cells by HF treatment of the oxide layer.* Applied Physics Letters, 1976. **28**(10): p. 620-621.

159. Shewchun, J., et al. *Photovoltaic characteristic of tunnel MIS silicon solar cells.* in *Proc. of the National Workshop on Low Cost Polycrystalline Silicon Solar Cells.* 1976. Southern Methodist University, Dallas, Texas

160. Feteha, M.Y., et al., *Metal-insulator-semiconductor solar cell under gamma irradiation.* Renewable Energy, 2002. **26**(1): p. 113-120.

161. Feteha, M.Y.M., *Heterojunction AlGaAs-GaAs solar cells for space applications,* in *Electrical and Electronic Engineering.* 1995, University of Central Lancashire. p. 280.

162. So, S.M., et al., *Collection efficiency of thin-film metal-insulator-semiconductor solar cells.* Journal of Applied Physics, 1984. **55**(1): p. 253-261.

163. Al-Mass'ari,   M.A.S.,   *Computer   modelling   of   an   Al/SiO2/Si metal/insulator/semiconductor solar cell.* Solar Cells, 1985. **14**(2): p. 99-108.

164. Divigalpitiya, W.M.R. and S.R. Morrison, *The dependence of photovoltaic characteristics of metal-insulator-semiconductor solar cells on dislocation density.* Journal of Applied Physics, 1989. **66**(9): p. 4462-4465.

165. Sen, K. and R.S. Srivastava, *Variation in the diode quality factor for metal-insulator-n-semiconductor solar cells.* Solar Cells, 1982. **7**(3): p. 219-223.

166. Goel, A. and T.P. Sharma, *Change in the diode quality factor with insulator layer thickness in a metal-insulator-n-semiconductor solar cell.* Journal of Applied Physics, 1985. **57**(8): p. 2973-2974.

167. Wong, D.Y.F. and Y.W. Lam, *Preprocessing heat treatment of metal-insulator-semiconductor solar cells.* Applied Physics Letters, 1986. **48**(15): p. 981-982.

168. Daliakouras, N., *Automatic creation of questions / exercises for an artificial intelligence teaching system.* 1984.

169. Babsail, L.S. and S.R. Morrison, *The effect of hydrogen treatment on damaged and undamaged metal-insulator-semiconductor solar cells.* Journal of Applied Physics, 1991. **70**(1): p. 259-265.

170. Gomaa, N.G., *Photon-induced degradation in metal–insulator–semiconductor solar cells.* Renewable Energy, 2001. **24**(3): p. 529-534.

171. Grimshaw, J.A. and W.G. Townsend, *Photon-induced degradation in crystalline silicon Schottky barrier solar cells.* Solar Cells, 1980. **2**(1): p. 55-63.

172. Soliman, M.M., *Degradation of transparent metal–insulator–semiconductor solar cells due to heating effects.* Renewable Energy, 2001. **23**(3): p. 483-488.

173. Cheek, G. and R. Mertens, *Metal-insulator-semiconductor silicon solar cells.* Solar Cells, 1983. **8**(1): p. 17-32.

174. Platzer-Björkman, C., *Kesterite compound semiconductors for thin film solar cells.* Current Opinion in Green and Sustainable Chemistry, 2017. **4**: p. 84-90.

175. Rajeswaran, G., et al., *A stable ytterbium-insulator-semiconductor solar cell based on an interface degradation model.* IEEE Transactions on Electron Devices, 1983. **30**(12): p. 1840-1842.

176. Anderson, W.A., et al., *A revised process to increase efficiency and reproducibility in Cr-MIS solar cells.* IEEE Electron Device Letters, 1980. **1**(7): p. 128-130.

177. Fathi, M. and A. Mougas, *Recycling of a defective metal insulator semiconductor solar cell by hot phosphoric acid.* Journal of Renewable and Sustainable Energy, 2011. **3**(1): p. 013103.

178. Har-Lavan, R., et al. *Hybrid, chemically passivated n-type silicon/PEDOT:PSS semiconductor-insulator-semiconductor solar cell.* in *2011 37th IEEE Photovoltaic Specialists Conference.* 2011.

179. Shewchun, J., et al., *The operation of the semiconductor-insulator-semiconductor (SIS) solar cell: Theory.* Journal of Applied Physics, 1978. **49**(2): p. 855-864.

180. Oakley, R.E. and G.A. Godber, *Growth of very thin oxide films on silicon for use in MNOS charge storage devices.* Thin Solid Films, 1972. **9**(2): p. 287-291.

181. Goodman, A.M. and J.M. Breece, *Thin Tunnelable Layers of Silicon Dioxide Formed by Oxidation of Silicon.* Journal of The Electrochemical Society, 1970. **117**(7): p. 982.

182. Fehlner, F.P., *Formation of Ultrathin Oxide Films on Silicon.* Journal of The Electrochemical Society, 1972. **119**(12): p. 1723.

183. Saha, N.R., D. Roychaudhuri, and P.K. Basu, *Barrier height enhancement in semiconductor-insulator-semiconductor solar cells due to surface states and insulator charges.* Solar Cells, 1983. **8**(4): p. 397-401.

184. Shewchun, J., D. Burk, and M.B. Spitzer, *MIS and SIS solar cells.* IEEE Transactions on Electron Devices, 1980. **27**(4): p. 705-716.

185. Ashok, S., P.P. Sharma, and S.J. Fonash, *Spray-deposited ITO—Silicon SIS heterojunction solar cells.* IEEE Transactions on Electron Devices, 1980. **27**(4): p. 725-730.

186. Calderer, J., et al., *Caractérisation des cellules solaires silicium (n)-In2O3 (dope Sn) préparées par une méthode de vaporisation.* Rev. Phys. Appl. (Paris), 1979. **14**(3): p. 485-490.

187. Wager, J.F. and C.W. Wilmsen, *Detection of SiO2 at the indium tin oxide/Si solar cell interface.* Journal of Applied Physics, 1979. **50**(6): p. 4172-4177.

188. Vishwakarma, S.R., Rahmatullah, and H.C. Prasad, *Fabrication of SnO2:As/SiO2/n-Si (textured) (semiconductor/insulator/semiconductor) solar cells by chemical vapor deposition.* Journal of Applied Physics, 1991. **70**(12): p. 7474-7477.

189. Roy, R. and M.W. Shafer, *Phases Present and Phase Equilibrium in the System In2O3–H2O.* The Journal of Physical Chemistry, 1954. **58**(4): p. 372-375.

190. Lehmann, H.W. and R. Widmer, *Preparation and properties of reactively co-sputtered transparent conducting films.* Thin Solid Films, 1975. **27**(2): p. 359-368.

191. Fan, J.C.C. and J.B. Goodenough, *X-ray photoemission spectroscopy studies of Sn-doped indium-oxide films.* Journal of Applied Physics, 1977. **48**(8): p. 3524-3531.

192. Schunck, J.P. and A. Coche, *Efficient indium tin oxide/polycrystalline silicon solar cells.* Applied Physics Letters, 1979. **35**(11): p. 863-865.

193. Ghosh, A.K., C. Fishman, and T. Feng, *Theory of the electrical and photovoltaic properties of polycrystalline silicon.* Journal of Applied Physics, 1980. **51**(1): p. 446-454.

194. Smith, P., R. Singh, and J. DuBow, *Reverse current-voltage characteristics of indium tin oxide/silicon solar cells under illumination.* Journal of Applied Physics, 1980. **51**(4): p. 2164-2166.

195. Oshima, T., et al., *Effects of SiH2Cl2 on low-temperature (≤200°C) Si epitaxy by photochemical vapor deposition.* Applied Surface Science, 1994. **79-80**: p. 215-219.

196. Spitzer, M., J. Shewchun, and D. Burk, *The operation of the semiconductor-insulator-semiconductor solar cell: Barrier height lowering through interface states.* Journal of Applied Physics, 1980. **51**(12): p. 6399-6404.

197. Stirn, R.J. and Y.C.M. Yeh. *Single crystal and polycrystalline GaAs solar cells using AMOS technology.* in *Proc. 12th Photovoltaic Specialists Conf.* 1976. Baton Rouge, LA, New York.

198. Har-Lavan, R., et al., *Toward metal-organic insulator-semiconductor solar cells, based on molecular monolayer self-assembly on n-Si.* Applied Physics Letters, 2009. **94**(4): p. 043308.

199. Shaban, M. and M.A.-G. El-Sayed, *Temperature dependence characterization of metal–insulator–nonuniformly-doped semiconductor solar cell.* Applied Surface Science, 2008. **254**(23): p. 7901-7904.

200. Chen, T., et al., *Metal–insulator transition in films of doped semiconductor nanocrystals.* Nature Materials, 2016. **15**(3): p. 299-303.

201. Oener, S.Z., et al., *Metal–Insulator–Semiconductor Nanowire Network Solar Cells.* Nano Letters, 2016. **16**(6): p. 3689-3695.

202. Möller, H.J., *New Materials: Semiconductors for Solar Cells.* Handbook of Semiconductor Technology Set, 2000: p. 715-769.

203. Yamaguchi, M., *Radiation resistance of compound semiconductor solar cells.* Journal of Applied Physics, 1995. **78**(3): p. 1476-1480.

204. Walukiewicz, W., *Narrow band gap group III-nitride alloys.* Physica E: Low-dimensional Systems and Nanostructures, 2004. **20**(3): p. 300-307.

205. Schock, H.W. *Polycrystalline Compound Semiconductor Thin Films in Solar Cells.* in *Polycrystalline Semiconductors.* 1989. Berlin, Heidelberg: Springer Berlin Heidelberg.

206. Jeong, Y., et al., *III-V Tandem, CuInGa(S,Se)2, and Cu2ZnSn(S,Se)4 Compound Semiconductor Thin Film Solar Cells.* The Korean Society of Industrial and Engineering Chemistry, 2015. **26**(5): p. 526-532.

207. Sato, K. *Compound semiconductor-based solar cells.* in *4th International seminar for special doctoral program.* 2015. Yamanashi University.

208. Kim, M., et al., *Effect of the Impurity Incorporation on the Performance of Cu(In,Ga)Se2 Semiconductor Solar Cells.* Journal of Nanoscience and Nanotechnology, 2016. **16**(10): p. 10748-10752.

209. Siebentritt, S., *Chalcopyrite compound semiconductors for thin film solar cells.* Current Opinion in Green and Sustainable Chemistry, 2017. **4**: p. 1-7.

210. Wu, J., et al., *Narrow bandgap group III-nitride alloys.* physica status solidi (b), 2003. **240**(2): p. 412-416.

211. Anani, M., et al., *High-grade efficiency III-nitrides semiconductor solar cell.* Microelectronics Journal, 2009. **40**(3): p. 427-434.

212. Zekry, A., et al., *Capacitance and conductance of Zn/sub x/Cd/sub 1-x/S/ZnTe heterojunctions.* IEEE Transactions on Electron Devices, 1993. **40**(2): p. 259-266.

213. Jongseung, Y. *Ultrathin, microscale epitaxial compound semiconductor solar cells.* in *Proc.SPIE.* 2011.

214. Oh, S., et al., *Compound semiconductor solar cells on Si substrate for medium concentrator photovoltaic applications.* AIP Conference Proceedings, 2013. **1556**(1): p. 101-105.

215. Rayimjon, A., I.R. Gulomjonovich, and A.M. Alisherovna, *Temperature Stimulation of Effective Value of Density of the Current of Semiconductor Solar Cells.* Energy and Power Engineering, 2019. **11**(2): p. 92-97.

216. Inguanta, R., et al. *An electrochemical route towards the fabrication of nanostructured semiconductor solar cells.* in *SPEEDAM 2010.* 2010.

217. McGinnis, S.P. and B. Das. *A novel technique for fabricating semiconductor nanodevice arrays on silicon.* in *Proceedings IEEE/Cornell Conference on Advanced Concepts in High Speed Semiconductor Devices and Circuits.* 1995.

218. Haque, S.A., et al., *Flexible dye sensitised nanocrystalline semiconductor solar cells.* Chemical Communications, 2003(24): p. 3008-3009.

219. Kouhnavard, M., et al., *A review of semiconductor materials as sensitizers for quantum dot-sensitized solar cells.* Renewable and Sustainable Energy Reviews, 2014. **37**: p. 397-407.

220. Prakash, T., *Review on nanostructured semiconductors for dye sensitized solar cells.* Electronic Materials Letters, 2012. **8**(3): p. 231-243.

221. Yoshioka, S., T. Mishima, and M. Ihara, *The Effect of TiO2 Microstructure and Introduction of Silver Nanoparticles on Conversion Efficiency of Sb2S3 Sensitized Semiconductor Solar Cells.* ECS Transactions, 2013. **50**(51): p. 33-44.

222. Termsaithong, P. and A. Tubtimtae, *Boron-doped CuInTe2 semiconductor-sensitized liquid-junction solar cells.* Materials Letters, 2015. **138**: p. 41-44.

223. Huang, P.-C., W.-C. Yang, and M.-W. Lee, *AgBiS2 Semiconductor-Sensitized Solar Cells.* The Journal of Physical Chemistry C, 2013. **117**(36): p. 18308-18314.

224. Tubtimtae, A., et al., *Tailoring CuxS semiconductor-sensitized SnO2 solar cells.* Materials Letters, 2015. **147**: p. 16-19.

225. Milanesio, M.E., et al., *Synthesis of a diporphyrin dyad bearing electron-donor and electron-withdrawing substituents with potential use in the spectral sensitization of semiconductor solar cells.* Journal of Porphyrins and Phthalocyanines, 2003. **07**(01): p. 42-51.

226. Chen, C., et al. *Evaluation of Luminescent Downshifting Effect of Perovskite Quantum Dots on Semiconductor Solar Cells.* in *2018 IEEE 7th World Conference on Photovoltaic Energy Conversion (WCPEC) (A Joint Conference of 45th IEEE PVSC, 28th PVSEC & 34th EU PVSEC).* 2018.

227. Fukui, T., et al. *III-V Compound Semiconductor Nanowire Solar Cells.* in *CLEO: 2013.* 2013. San Jose, California: Optical Society of America.

228. Routkevitch, D., et al., *Nonlithographic nano-wire arrays: fabrication, physics, and device applications.* IEEE Transactions on Electron Devices, 1996. **43**(10): p. 1646-1658.

229. Thony, P., *15 - Semiconductor nanowires for solar cells*, in *Semiconductor Nanowires*, J. Arbiol and Q. Xiong, Editors. 2015, Woodhead Publishing. p. 411-439.

230. Guo, W., Z. Xu, and T. Li, *8 - Metal-based semiconductor nanomaterials for thin-film solar cells*, in *Multifunctional Photocatalytic Materials for Energy*, Z. Lin, M. Ye, and M. Wang, Editors. 2018, Woodhead Publishing. p. 153-185.

231. Zunger, A., S. Wagner, and P.M. Petroff, *New materials and structures for photovoltaics.* Journal of Electronic Materials, 1993. **22**(1): p. 3-16.

232. Cavallo, C., et al., *Nanostructured Semiconductor Materials for Dye-Sensitized Solar Cells.* Journal of Nanomaterials, 2017. **2017**: p. 5323164.

233. Lira-Cantu, M., *The future of semiconductor oxides in next-generation solar cells.* 2018: Elsevier.

234. Zhang, W.-H. and B. Cai, *Organolead halide perovskites: a family of promising semiconductor materials for solar cells.* Chinese Science Bulletin, 2014. **59**(18): p. 2092-2101.

235. Yang, Y., et al., *Inorganic p-type semiconductors and carbon materials based hole transport materials for perovskite solar cells.* Chinese Chemical Letters, 2018. **29**(8): p. 1242-1250.

236. Srivastava, G. and R. Kumar, *Organic semiconductors and structural properties of tandem solar cell.* International Journal of Engineering and Advanced Technology (IJEAT), 2019. **9**: p. 532-534.

237. Borrelli, N.F., et al., *Quantum confinement effects of semiconducting microcrystallites in glass.* Journal of Applied Physics, 1987. **61**(12): p. 5399-5409.

238. Itoh, T. and T. Kirihara, *Excitons in CuCl microcrystals embedded in NaCl.* Journal of Luminescence, 1984. **31-32**: p. 120-122.

239. Leon, R., et al., *Spatially Resolved Visible Luminescence of Self-Assembled Semiconductor Quantum Dots.* Science, 1995. **267**(5206): p. 1966.

240. Ozin, G.A., *The zeolate ligand: From hydrolysis to capped semiconductor nanoclusters.* Advanced Materials, 1994. **6**(1): p. 71-76.

241. Martin, C.R., *Membrane-Based Synthesis of Nanomaterials.* Chemistry of Materials, 1996. **8**(8): p. 1739-1746.

242. Morgan, R.A., et al., *Experimental studies of the non-linear optical properties of cadmium selenide quantum-confined microcrystallites.* Semiconductor Science and Technology, 1990. **5**(6): p. 544-548.

243. Kurtz, S.R., D. Myers, and J.M. Olson. *Projected performance of three- and four-junction devices using GaAs and GaInP.* in *Conference Record of the Twenty Sixth IEEE Photovoltaic Specialists Conference - 1997.* 1997.

244. Vaquero-Stainer, A., et al., *Semiconductor nanostructure quantum ratchet for high efficiency solar cells.* Communications Physics, 2018. **1**(1): p. 7.

245. Martí, A., et al., *Novel semiconductor solar cell structures: The quantum dot intermediate band solar cell.* Thin Solid Films, 2006. **511-512**: p. 638-644.

246. Marti, A., L. Cuadra, and A. Luque. *Quantum dot intermediate band solar cell.* in *Conference Record of the Twenty-Eighth IEEE Photovoltaic Specialists Conference - 2000 (Cat. No.00CH37036).* 2000.

247. *Next Generation Photovoltaics: High Efficiency through Full Spectrum Utilization.* 1 ed. 2003: CRC Press.

248. Tian, J. and G. Cao, *Semiconductor quantum dot-sensitized solar cells.* Nano Reviews, 2013. **4**(1): p. 22578.

249. Ranjan, R., et al., *Chapter 7 - Metal and metal-semiconductor core–shell nanostructures for plasmonic solar cell applications*, in *Metal Semiconductor Core-Shell Nanostructures for Energy and Environmental Applications*, R.K. Gupta and M. Misra, Editors. 2017, Elsevier. p. 159-177.

250. Gupta, K.M. and N. Gupta, *Special Semiconducting Materials in Vivid Fields (for Thermoelectrics, Integrated Circuits, Photocatalytics, Spintronic Devices, etc.), Plasmonic Solar Cell, and Photonics*, in *Advanced Semiconducting Materials and Devices*, K.M. Gupta and N. Gupta, Editors. 2016, Springer International Publishing: Cham. p. 477-507.

251. Mokkapati, S., et al. *Plasmonics for III–V semiconductor solar cells*. in *IEEE Photonics Conference 2012*. 2012.

252. Li, Z., et al., *III–V Semiconductor Single Nanowire Solar Cells: A Review*. Advanced Materials Technologies, 2018. **3**(9): p. 1800005.

253. Sugaya, T., *MBE of III–V Semiconductors for Solar Cells*. Molecular Beam Epitaxy, 2019: p. 265-278.

254. Luque, A. and A.V. Mellor, *Photon Absorption Models in Nanostructured Semiconductor Solar Cells and Devices*. SpringerBriefs in Applied Sciences and Technology. 2015: Springer, Cham.

255. Ionescu, D. and M. Kovaci. *Performances of a metamaterial based on semiconductors for solar cell applications*. in *2018 International Symposium on Electronics and Telecommunications (ISETC)*. 2018.

256. Wu, J., et al., *Multi-type quantum dots photo-induced doping enhanced graphene/semiconductor solar cell*. RSC Advances, 2017. **7**(53): p. 33413-33418.

257. Szostak, R., et al., *Chapter 10 - Application of Graphene and Graphene Derivatives/Oxide Nanomaterials for Solar Cells*, in *The Future of Semiconductor Oxides in Next-Generation Solar Cells*, M. Lira-Cantu, Editor. 2018, Elsevier. p. 395-437.

258. Basyoni, M.S.S., A. Zekry, and A. Shaker, *Investigation of Base High Doping Impact on the npn Solar Cell Microstructure Performance Using Physically Based Analytical Model*. IEEE Access, 2021. **9**: p. 16958-16966.

259. Salem, M.S., et al., *Physically Based Analytical Model of Heavily Doped Silicon Wafers Based Proposed Solar Cell Microstructure*. IEEE Access, 2020. **8**: p. 138898-138906.

260. Salem, M.S., et al. *Design and simulation of proposed low cost solar cell structures based on heavily doped silicon wafers*. in *2016 IEEE 43rd Photovoltaic Specialists Conference (PVSC)*. 2016.

261. Salem, M.S., et al. *Effect of base width variation on the performance of a proposed ultraviolet low cost high efficiency solar cell structure*. in *2012 38th IEEE Photovoltaic Specialists Conference*. 2012.

262. Salem, M.S., et al., *Performance enhancement of a proposed solar cell microstructure based on heavily doped silicon wafers*. Semiconductor Science and Technology, 2019. **34**(3): p. 035012.

263. Rao, Z., et al., *Revisit of amorphous semiconductor InGaZnO4: A new electron transport material for perovskite solar cells.* Journal of Alloys and Compounds, 2019. **789**: p. 276-281.

264. Spear, W.E. and P.G. Le Comber, *Substitutional doping of amorphous silicon.* Solid State Communications, 1975. **17**(9): p. 1193-1196.

265. Fritzsche, H., C.C. Tsai, and P. Persans, *AMORPHOUS SEMICONDUCTING SILICON-HYDROGEN ALLOYS.* 1978.

266. Carlson, D.E., *Amorphous silicon solar cells.* IEEE Transactions on Electron Devices, 1977. **24**(4): p. 449-453.

267. Wronski, C.R., *Chapter 10 The Staebler-Wronski Effect*, in *Semiconductors and Semimetals*, J.I. Pankove, Editor. 1984, Elsevier. p. 347-374.

268. Goodman, C.H.L., *Improved amorphous semiconductors for solar cells.* Nature, 1979. **279**(5711): p. 349-349.

269. Lopez-Varo, P., et al., *Physical aspects of ferroelectric semiconductors for photovoltaic solar energy conversion.* Physics Reports, 2016. **653**: p. 1-40.

270. Wolf, M., *A new look at silicon solar cell performance.* Energy Conversion, 1971. **11**(2): p. 63-73.

271. Zekry, Z. and G. Eldallal, *Effect of MS contact on the electrical behaviour of solar cells.* Solid-State Electronics, 1988. **31**(1): p. 91-97.

272. Ghahremani, A. and A.E. Fathy, *A three-dimensional multiphysics modeling of thin-film amorphous silicon solar cells.* Energy Science & Engineering, 2015. **3**(6): p. 520-534.

273. Burgelman, M., P. Nollet, and S. Degrave, *Modelling polycrystalline semiconductor solar cells.* Thin Solid Films, 2000. **361-362**: p. 527-532.

274. Burgelman, M. and B. Minnaert, *Including excitons in semiconductor solar cell modelling.* Thin Solid Films, 2006. **511-512**: p. 214-218.

275. Islam, M.J., et al. *Modeling of graphene/SiO2/Si(n) based metal-insulator-semiconductor solar cells.* in *2016 4th International Conference on the Development in the in Renewable Energy Technology (ICDRET).* 2016.

276. Paul, P., B.W. Jonathan, and R.B. George. *Modeling, synthesis, and testing of materials and devices for organic semiconductor solar cells.* in *Proc.SPIE.* 1992.

277. Werner, J.H., R. Brendel, and H.J. Queisser. *New upper efficiency limits for semiconductor solar cells.* in *Proceedings of 1994 IEEE 1st World Conference on Photovoltaic Energy Conversion - WCPEC (A Joint Conference of PVSC, PVSEC and PSEC).* 1994.

278. Brendel, R., J.H. Werner, and H.J. Queisser, *Thermodynamic efficiency limits for semiconductor solar cells with carrier multiplication.* Solar Energy Materials and Solar Cells, 1996. **41-42**: p. 419-425.

279. Ding, D., et al., *A semi-analytical model for semiconductor solar cells.* Journal of Applied Physics, 2011. **110**(12): p. 123104.

280. Xu, T., et al., *Strategic improvement of the long-term stability of perovskite materials and perovskite solar cells.* Physical Chemistry Chemical Physics, 2016. **18**(39): p. 27026-27050.

Semiconductor Materials and Modelling for Solar Cells   Materials Research Forum LLC
Materials Research Foundations **104** (2021)   https://doi.org/10.21741/9781644901434

281. Olsen, L.C., *Model calculations for metal-insulator-semiconductor solar cells.* Solid-State Electronics, 1977. **20**(9): p. 741-751.

Semiconductor Materials and Modelling for Solar Cells   Materials Research Forum LLC
Materials Research Foundations **104** (2021)   https://doi.org/10.21741/9781644901434

Keyword Index

About the Authors

Eng. Zahra Pezeshki is a master's holder in Electrical Engineering-Electronics Integrated Circuits from the Shahrood University of Technology, Faculty of Electrical and Robotics Engineering where she presented a new method for energy consumption improvement by determination of the best location for cooling and heating appliances in buildings. She has worked as a lecturer, instructor, and adjunct professor with companies, academies, and universities. She has reviewed for 23 double-blind peer-review international journals and 25 international conferences and published many papers over national and international journals, and conference proceedings as well as book chapters. She has won numerous scientific awards and grants from renowned academic bodies and filled a patent on irrigation. She has worked as an invited researcher with national and international universities. Her broad research interests cover topics relating to electrical and electronics, computer, energy, mechanics, chemistry, civil, and physics. Zahra Pezeshki is an editorial board member of 9 international journals. In addition, she is a member of scientific organizations and over the last 19 years has built strong working collaborations with reputable groups in numerous and leading universities and organizations in her country and across the globe. Overall, she has good research experience in multidisciplinary fields of computer, energy, mechanics, chemistry, civil, and physics, and, more specifically, renewable energy and environment, and also works on multiple key research projects funded by institutional and government agencies.

Dr. Abdelhalim Zekry is a full professor of electronics at faculty of Engineering, Ain Shams University, Department of Electronics and Electrical Communication Engineering, Egypt. He has worked as a staff member on several universities. He has published more than 300 papers. He has also supervised more than 110 Master thesis and 40 Doctorate. Prof. Zekry focuses his research programs on the field of microelectronics and electronic applications including communications and photovoltaics. He has got several prizes for his outstanding research and teaching performance.

About the Editor

Dr. Inamuddin is working as Assistant Professor at the Department of Applied Chemistry, Aligarh Muslim University, Aligarh, India. He obtained Master of Science degree in Organic Chemistry from Chaudhary Charan Singh (CCS) University, Meerut, India, in 2002. He received his Master of Philosophy and Doctor of Philosophy degrees in Applied Chemistry from Aligarh Muslim University (AMU), India, in 2004 and 2007, respectively. He has extensive research experience in multidisciplinary fields of Analytical Chemistry, Materials Chemistry, and Electrochemistry and, more specifically, Renewable Energy and Environment. He has worked on different research projects as project fellow and senior research fellow funded by University Grants Commission (UGC), Government of India, and Council of Scientific and Industrial Research (CSIR), Government of India. He has received Fast Track Young Scientist Award from the Department of Science and Technology, India, to work in the area of bending actuators and artificial muscles. He has completed four major research projects sanctioned by University Grant Commission, Department of Science and Technology, Council of Scientific and Industrial Research, and Council of Science and Technology, India. He has published 185 research articles in international journals of repute and nineteen book chapters in knowledge-based book editions published by renowned international publishers. He has published 120 edited books with Springer (U.K.), Elsevier, Nova Science Publishers, Inc. (U.S.A.), CRC Press Taylor & Francis Asia Pacific, Trans Tech Publications Ltd. (Switzerland), IntechOpen Limited (U.K.), Wiley-Scrivener, (U.S.A.) and Materials Research Forum LLC (U.S.A). He is a member of various journals' editorial boards. He is also serving as Associate Editor for journals (Environmental Chemistry Letter, Applied Water Science and Euro-Mediterranean Journal for Environmental Integration, Springer-Nature), Frontiers Section Editor (Current Analytical Chemistry, Bentham Science Publishers), Editorial Board Member (Scientific Reports-Nature), Editor (Eurasian Journal of Analytical Chemistry), and Review Editor (Frontiers in Chemistry, Frontiers, U.K.) He is also guest-editing various special thematic special issues to the journals of Elsevier, Bentham Science Publishers, and John Wiley & Sons, Inc. He has attended as well as chaired sessions in various international and national conferences. He has worked as a Postdoctoral Fellow, leading a research team at the Creative Research Initiative Center for Bio-Artificial Muscle, Hanyang University, South Korea, in the field of renewable energy, especially biofuel cells. He has also worked as a Postdoctoral Fellow at the Center of Research Excellence in Renewable Energy, King Fahd University of Petroleum and Minerals, Saudi Arabia, in the field of polymer electrolyte membrane fuel cells and computational fluid dynamics of polymer electrolyte membrane fuel cells. He is a life member of the Journal of the Indian Chemical Society. His research interest includes ion exchange materials, a sensor for heavy metal ions, biofuel cells, supercapacitors and bending actuators.

www.ingramcontent.com/pod-product-compliance
Lightning Source LLC
Chambersburg PA
CBHW071500210326
41597CB00018B/2634